AF572781

Biotech's

DICTIONARY *Of* ENERGY

Compiled & Edited by :

JB Jain & Sumit Jain

2006

BIOTECH BOOKS

DELHI - 110035

Biotech's

DICTIONARY *Of* ENERGY

Compiled & Edited by :
JB Jain & Sumit Jain

First Indian Edition 2006

ISBN 81-7622-165-1

Published by : BIOTECH BOOKS
1123/74, Tri Nagar,
Delhi - 110 035
Phone: 27382765
e-mail: biotechbooks@yahoo.co.in

Showroom : 4762-63/23, Ansari Road, Darya Ganj,
New Delhi - 110 002
Phone: 23245578, 23244987

Printed at : Chawla Offset Printers
New Delhi - 110 052

PRINTED IN INDIA

PREFACE

Energy resources vary widely in technical and economic characteristics. Some resources, such as wind, geothermal, modern biomass, and small hydroelectric energy, are in fairly wide use throughout the world, are often economical, and offer significant environmental advantages. Those energy resources are applicable for either grid use or for stand-alone energy in rural communities. Other renewable resources, such as photovoltaics, remain too expensive for many electric grid applications, but are well suited for grid niche applications, such as for switching equipment upgrades. For poor and remote communities not yet served by electricity, the above-mentioned renewable resources are highly economical, particularly to provide power for lighting, refrigeration, irrigation, and communications. (2) In addition, modern biomass applications are particularly advantageous for developing countries because they use local feed stocks and labour. (3) Other renewable resources with tremendous technical and economic potential such as hydrogen fuel cells, wave and tidal energy, and deep hot rock geothermal energy, require additional research and development to be economically or technically feasible. (4)

Nuclear energy is excluded from this analysis as a development option because of its high capital and operating costs, complex technical requirements for operation and maintenance, and unresolved problems of proliferation and waste disposal.

In addition, nuclear energy is derived from plutonium or uranium processed with high energy into forms capable of utilization in reactors. If fossil fuels are used as the energy source to refine the uranium (currently the usual process) then nuclear energy has much of the same carbon dioxide and pollution problems as direct fossil fuel combustion. (5) Nuclear power waste disposal and plant decommissioning also involve substantial unsolved environmental problems and costs. Finally, there are safety problems with nuclear power plant operations, and risks of diversion of nuclear fuel to weapons production. Because nuclear power requires safety and high capital costs for construction, waste disposal and decommissioning.

It is obviously a matter of great urgency that all countries take effective measures to reduce their energy consumption, but no country can afford to wait for a unified global effort. For the developing countries it is crucial both to continue and to expand efforts to use energy efficiently and to increase production. They have little margin to tolerate waste, their energy requirements are growing rapidly, and the achievement of their long-term development objectives depends on accommodating to the new energy situation. Although their consumption of commercial energy is relatively small, they are in a sense better placed to adjust their use of it than the industrialized countries, in that they are less committed to a capital stock and life style evolved in an era of cheap coal and oil. But they are hampered by indadequate information about resources and uses. Most of them also lack experience in commercial fuel production. The large majority have not yet tapped their own resources to any considerable degree and there is great scope for reducing dependence on imports.

In designing policies to help resolve its energy problems, every country faces a unique set of conditions, including its level of income and degree of industrialization, its energy resource endowment, the relative importance of commercial and traditional fuels, its degree of dependence on oil imports, and other factors. But since the rise in the price of petroleum, the degree of dependence on petroleum imports has become the most important single factor.

The Dictionary of Energy resources is an authoritative and wide-ranging reference book covering all aspects of energy and its role in society. It integrates all of the social, natural and engineering sciences alongside a wide range of entirely new areas, such as: energy and sustainable development; energy flows in the biological realm; methods of energy modelling and accounting; energy and materials; and Net Energy Analysis.

It is an essential reference companion for anyone with an interest in energy, including scientists, consultants, journalists, economists and policy-makers.

acbm

Acronym for "asbestos-containing building material."

account classification

The way in which suppliers of electricity, natural gas, or fuel oil classify and bill their customers. Commonly used account classifications are "Residential," "Commercial," "Industrial," and "Other." Suppliers' definitions of these terms vary from supplier to supplier. In addition, the same customer may be classified differently by each of its energy suppliers.

account of others (natural gas)

Natural gas deliveries for the account of others are deliveries to customers by transporters that do not own the natural gas but deliver it for others for a fee. Included are quantities covered by long-term contracts and quantities involved in short-term or spot market sales.

accounting system

A method of recording accounting data for a utility or company or a method of supplying accounting information for controlling, evaluating, planning and decision making.

acid mine drainage

This refers to water pollution that results when sulfur-bearing minerals associated with coal are exposed to air and water and form sulfuric acid and ferrous sulfate. The ferrous sulfate can further react to form ferric hydroxide, or yellow boy, a yellow-orange iron precipitate found in streams and rivers polluted by acid mine drainage.

acid rain

Also called acid precipitation or acid deposition, acid rain is precipitation containing harmful amounts of nitric and sulfuric acids formed primarily by sulfur dioxide and nitrogen oxides released into the atmosphere when fossil fuels are burned. It can be wet precipitation (rain, snow, or fog) or dry precipitation (absorbed gaseous and particulate matter, aerosol particles or dust). Acid rain has a pH below 5.6. Normal rain has a pH of about 5.6, which is slightly acidic. The term pH is a measure of acidity or alkalinity and ranges from 0 to 14. A pH measurement of 7 is regarded as neutral. Measurements below 7 indicate increased acidity, while those above indicate increased alkalinity.

acquisition (foreign crude oil)

All transfers of ownership of foreign crude oil to a firm, irrespective of the terms of that transfer. Acquisitions thus include all purchases and exchange receipts as well as any and all foreign crude acquired under reciprocal buy-sell agreements or acquired as a result of a buy-back or other preferential agreement with a host government.

- **acquisition (minerals)**
  The procurement of the legal right to explore for and produce discovered minerals, if any, within a specific area; that legal right may be obtained by mineral lease, concession, or purchase of land and mineral rights or of mineral rights alone.

- **acquisition costs, mineral rights**
  Direct and indirect costs incurred to acquire legal rights to extract natural resources. Direct costs include costs incurred to obtain options to lease or purchase mineral rights and costs incurred for the actual leasing (e.g., lease bonuses) or purchasing of the rights. Indirect costs include such costs as brokers' commissions and expenses; abstract and recording fees; filing and patenting fees; and costs for legal examination of title and documents.

- **acreage**
  An area, measured in acres, that is subject to ownership or control by those holding total or fractional shares of working interests. Acreage is considered developed when development has been completed. A distinction may be made between "gross" acreage and "net" acreage:
  Gross - All acreage covered by any working interest, regardless of the percentage of ownership in the interest.
  Net - Gross acreage adjusted to reflect the percentage of ownership in the working interest in the acreage.

- **acre-foot**
  The volume of water that will cover an area of 1 acre to a depth of 1 foot.

- **active solar**
  As an energy source, energy from the sun collected and stored using mechanical pumps or fans to circulate heat-laden fluids or air between solar collectors and a building.

- **actual peak reduction**
  The actual reduction in annual peak load (measured in kilowatts) achieved by customers that participate in a utility demand-side management (DSM) program. It reflects the changes in the demand for electricity resulting from a utility DSM program that is in effect at the same time the utility experiences its annual peak load, as opposed to the installed peak load reduction capability (i.e., potential peak reduction). It should account for the regular cycling of energy efficient units during the period of annual peak load.

adjustable speed drives

Drives that save energy by ensuring the motor's speed is properly matched to the load placed on the motor. Terms used to describe this category include polyphase motors, motor over sizing, and motor rewinding.

adjusted electricity

A measurement of electricity that includes the approximate amount of energy used to generate electricity. To approximate the adjusted amount of electricity, the site-value of the electricity is multiplied by a factor of 3. This conversion factor of 3 is a rough approximation of the Btu value of raw fuels used to generate electricity in a steam-generation power plant.

adjustment bid

A bid auction conducted by the independent system operator or power exchange to redirect supply or demand of electricity when congestion is anticipated.

administrative and general expenses

Expenses of an electric utility relating to the overall directions of its corporate offices and administrative affairs, as contrasted with expenses incurred for specialized functions. Examples include office salaries, office supplies, advertising, and other general expenses.

advance ro yalty

A royalty required to be paid in advance of production from a mineral property that may or may not be recoverable from future production.

advances from municipality

The amount of loans and advances made by the municipality or its other departments to the utility department when suc bject to current settlement.

advances to municipality

The amount of loans and advances made by the utility department to the municipality or its other departments when such loans or advances are subject to current settlement.

adverse water conditions

Reduced stream flow, lack of rain in the drainage basin, or low water supply behind a pondage or reservoir dam resulting in a reduced gross head that limits the production of hydroelectric power or forces restrictions to be placed on multipurpose reservoirs or other water uses.

affiliate

An entity which is directly or indirectly owned, operated, or controlled by another entity.

afforestation

Planting of new forests on lands that have not been recently forested.

▪ **aftermarket converted vehicle**
A standard conventionally fueled, factory-produced vehicle to which equipment has been added that enables the vehicle to operate on alternative fuel.

▪ **aftermarket vehicle converter**
An organization or individual that modifies OEM vehicles after first use or sale to operate on a different fuel (or fuels).

▪ **agglomerating character**
Agglomeration describes the caking properties of coal. Agglomerating character is determined by examination and testing of the residue when a small powdered sample is heated to 950 degrees Centigrade under specific conditions. If the sample is "agglomerating," the residue will be coherent, show swelling or cell structure, and be capable of supporting a 500-gram weight without pulverizing.

▪ **agglomerating**
refers to coal that softens when heated and forms a hard gray coke; this coal is called caking coal. not all caking coals are coking coals. the agglomerating value is used to differentiate between coal ranks and also is a guide to determine how a particular coal reacts in a furnace.

▪ **agglutinating**
refers to the binding qualities of a coal. the agglutinating value is an indication of how well a coke made from a particular coal will perform in a blast furnace. it is also called a caking index.

▪ **aggregate ratio**
The ratio of two population aggregates (totals). For example, the aggregate expenditures per household is the ratio of the total expenditures in each category to the total number of households in the category.

▪ **aggregator**
Any marketer, broker, public agency, city, county, or special district that combines the loads of multiple end-use customers in negotiating the purchase of electricity, the transmission of electricity, and other related services for these customers.

▪ **agriculture**
An energy-consuming subsector of the industrial sector that consists of all facilities and equipment engaged in growing crops and raising animals.

▪ **agriculture, mining, and construction (consumer category)**
Companies engaged in agriculture, mining (other than coal mining), or construction industries.

air cleaner

A device using filters or electrostatic precipitators to remove indoor-air pollutants such as tobacco smoke, dust, and pollen. Most portable units are 40 watts when operated on low speed and 100 watts on high speed.

air collector

A medium-temperature collector used predominantly in space heating, utilizing pumped air as the heat-transfer medium.

air conditioning

Cooling and dehumidifying the air in an enclosed space by use of a refrigeration unit powered by electricity or natural gas. Note: Fans, blowers, and evaporative cooling systems ("swamp coolers") that are not connected to a refrigeration unit are excluded.

air conditioning intensity

The ratio of air-conditioning consumption or expenditures to square footage of cooled floor space and cooling degree-days (base 65 degrees F). This intensity provides a way of comparing different types of housing units and households by controlling for differences in housing unit size and weather conditions. The square footage of cooled floor space is equal to the product of the total square footage times the ratio of the number of rooms that could be cooled to the total number of rooms. If the entire housing unit is cooled, the cooled floor space is the same as the total floor space. The ratio is calculated on a weighted, aggregate basis according to this formula: air-conditioning intensity = btu for air conditioning/(cooled square feet * cooling degree-days)

air pollution abatement equipment

Equipment used to reduce or eliminate airborne pollutants, including particulate matter (dust, smoke, fly, ash, dirt, etc.), sulfur oxides, nitrogen oxides (NO_x), carbon monoxide, hydrocarbons, odors, and other pollutants. Examples of air pollution abatement structures and equipment include flue-gas particulate collectors, flue-gas desulfurization units and nitrogen oxide control devices.

alcohol

The family name of a group of organic chemical compounds composed of carbon, hydrogen, and oxygen. The series of molecules vary in chain length and are composed of a hydrocarbon plus a hydroxyl group; $CH(3)\text{-}(CH(2))_n\text{-}OH$ (e.g., methanol, ethanol, and tertiary butyl alcohol).

■ **alkylate**

The product of an alkylation reaction. It usually refers to the high-octane product from alkylation units. This alkylate is used in blending high octane gasoline.

■ **alkylation**

A refining process for chemically combining isobutane with olefin hydrocarbons (e.g., propylene, butylene) through the control of temperature and pressure in the presence of an acid catalyst, usually sulfuric acid or hydrofluoric acid. The product, alkylate, an isoparaffin, has high octane value and is blended with motor and aviation gasoline to improve the antiknock value of the fuel.

■ **all-electric home**

A residence in which electricity is used for the main source of energy for space heating, water heating, and cooking. Other fuels may be used for supplementary heating or other purposes.

■ **alternate energy source for primary heater**

The fuel that would be used in place of the usual main heating fuel if the building had to switch fuels.

■ **alternating current (ac)**

An electric current that reverses its direction at regularly recurring intervals.

■ **alternative energy source1**

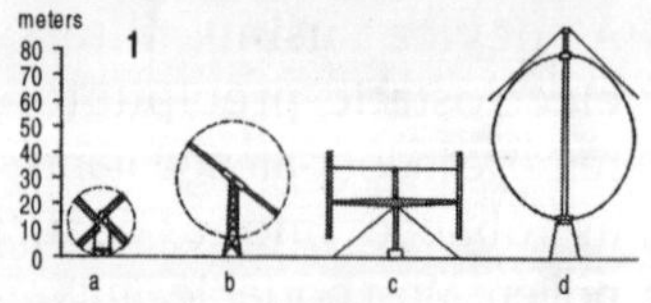

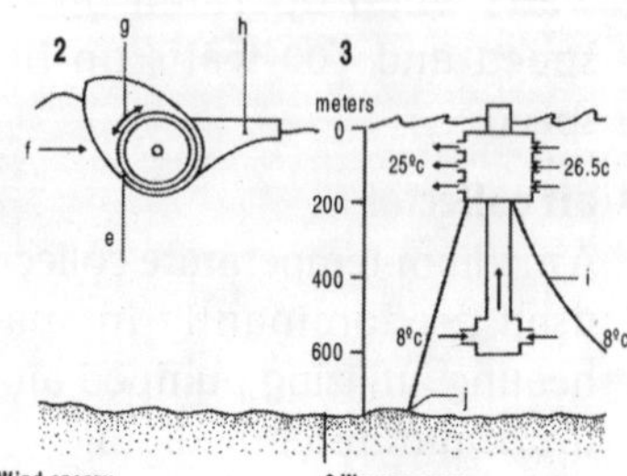

1 Wind energy
a Typical traditional windmill
b large horizontal axis wind turbine
c Large vertical axis wind -turbine (concept)
d Extra large vertical axis wind-turbine (concept)

2 Wave energy
e Rocking boom
f Wave
g Rocking motion
h Balancing float

3 Ocean thermal energy conversion (OTEC): general layout
i Power transmission line
j Anchor

■ **alternative energy source 2**

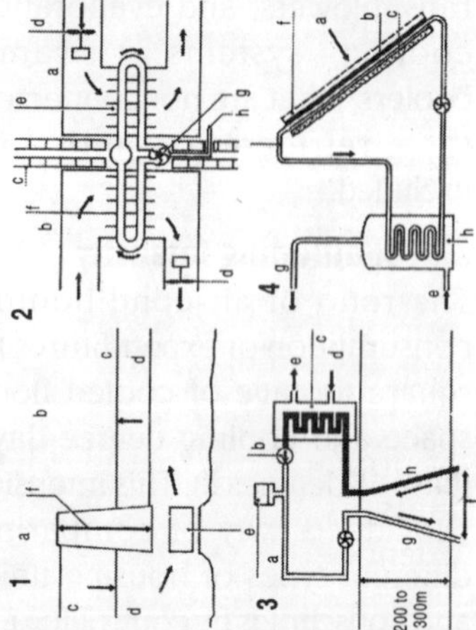

1 Tidal energy
a High basin
b Low basin
c Upper level
d Lower level
e Turbines

2 Heat pump
a Outside
b Inside
c House wall
d Fan
e Evaporator
f Condenser
g Compressor pump
h Electrical energy

3 Geothermal energy
a Expansion tank
b Flow control valve
c Heat exchange
f Permeable rock
g Cold water down
h Hot water up

4 Solar panel
a Sunlight
b Black heat absorbing surface Insulation
d Hot water
Cold water
f Transparent cover
g To domestic hot water system
h Panel heat exchange

■ **alternative fuel**

Alternative fuels, for transportation applications, include the following:

- methanol
- denatured ethanol, and other alcohols
- fuel mixtures containing 85 percent or more by volume of methanol, denatured ethanol, and other alcohols with gasoline or other fuels — natural gas
- liquefied petroleum gas (propane)
- hydrogen
- coal-derived liquid fuels
- fuels (other than alcohol) derived from biological materials (biofuels such as soy diesel fuel) electricity (including electricity from solar energy.)

"... any other fuel the Secretary determines, by rule, is substantially not petroleum and would yield substantial energy security benefits and substantial environmental benefits." The term "alternative fuel" does not include alcohol or other blended portions of primarily petroleum-based fuels used as oxygenates or extenders, i.e. MTBE, ETBE, other ethers, and the 10-percent ethanol portion of gasohol.

■ **alternative fuel vehicle converter**

An organization (including companies, government agencies and utilities), or individual that performs conversions involving alternative alternative fuel vehicles. An AFV converter can convert (1) conventionally fueled vehicles to AFVs, (2) AFVs to conventionally fueled vehicles, or (3) AFVs to use another alternative fuel.

■ **alternative-fuel vehicle (afv)**

A vehicle designed to operate on an alternative fuel (e.g., compressed natural gas, methane blend, electricity). The vehicle could be either a dedicated vehicle designed to operate exclusively on alternative fuel or a non dedicated vehicle designed to operate on alternative fuel and/ or a traditional fuel.

■ **alternative-rate dsm program assistance**

A DSM (demand-side management) program assistance that offers special rate structures or discounts on the consumer's monthly electric bill in exchange for participation in DSM programs aimed at cutting peak demands or changing load shape. These rates are intended to reduce consumer bills and shift hours of operation of equipment from on-peak to off-peak periods through the application of time-differentiated rates. For example, utilities often pay consumers several dollars a month (refund on their monthly electric bill) for participation in a load control program. Large commercial and industrial customers sometimes obtain interruptible rates, which provide a discount in return for the consumer's agreement to cut electric loads upon request from the utility (usually during critical periods, such as summer afternoons when the system demand approaches the utility's generating capability).

- **amorphous silicon**

An alloy of silica and hydrogen, with a disordered, non crystalline internal atomic arrangement, that can be deposited in thin-film layers (a few micrometers in thickness) by a number of deposition methods to produce thin-film photovoltaic cells on glass, metal, or plastic substrates.

- **amortization**

The depreciation, depletion, or charge-off to expense of intangible and tangible assets over a period of time. In the extractive industries, the term is most frequently applied to mean either (1) the periodic charge-off to expense of the costs associated with nonproducing mineral properties incurred prior to the time when they are developed and entered into production or (2) the systematic charge-off to expense of those costs of productive mineral properties (including tangible and intangible costs of prospecting, acquisition, exploration, and development) that had been initially capitalized (or deferred) prior to the time the properties entered into production, and thereafter are charged off as minerals are produced.

- **ampere**

The unit of measurement of electrical current produced in a circuit by 1 volt acting through a resistance of 1 Ohm.

- **ancillary services**

Services that ensure reliability and support the transmission of electricity from generation sites to customer loads. Such services may include: load regulation, spinning reserve, non-spinning reserve, replacement reserve, and voltage support.

- **annual operating factor**

The annual fuel consumption divided by the product of design firing rate and hours of operation per year.

- **annual requirement**

The reporting company's best estimate of the annual requirement for natural gas to make direct sales or sales for resale under certificate authorizations and for company use and unaccounted-for gas during the year next following the current report year.

ansi assembly identifier
The serial numbering scheme adopted by the American National Standards Institute (ANSI) to ensure uniqueness of an assembly serial number.

anthracite
The highest rank of coal; used primarily for residential and commercial space heating. It is a hard, brittle, and black lustrous coal, often referred to as hard coal, containing a high percentage of fixed carbon and a low percentage of volatile matter. The moisture content of fresh-mined anthracite generally is less than 15 percent. The heat content of anthracite ranges from 22 to 28 million Btu per ton on a moist, mineral-matter-free basis. The heat content of anthracite coal consumed in the United States averages 25 million Btu per ton, on the as-received basis (i.e., containing both inherent moisture and mineral matter).

anthropogenic
Made or generated by a human or caused by human activity. The term is used in the context of global climate change to refer to gaseous emissions that are the result of human activities, as well as other potentially climate-altering activities, such as deforestation.

api gravity
American Petroleum Institute measure of specific gravity of crude oil or condensate in degrees. An arbitrary scale expressing the gravity or density of liquid petroleum products. The measuring scale is calibrated in terms of degrees API; it is calculated as follows: Degrees API = (141.5 / sp.gr.60 deg.F/60 deg.F) - 131.5

api
The American Petroleum Institute, a trade association.

apparent consumption, (coal)
Coal production plus imports of coal, coke, and briquets minus exports of coal, coke, and briquets plus or minus stock changes. Note: The sum of "Production" and "Imports" less "Exports" may not equal "Consumption" due to changes in stocks, losses, unaccounted-for coal, and special arrangements such as the United States shipments of anthracite to United States Armed Forces in Europe.

apparent consumption, natural gas (international)
The total of an individual nation's dry natural gas production plus imports less exports.

apparent consumption, petroleum (international)
Consumption that includes internal consumption, refinery fuel and loss, and bunkering. For countries in the Organization for Economic Cooperation and

Development (OECD), apparent consumption is derived from refined product output plus refined product imports minus refined product exports plus refined product stock changes plus other oil consumption (such as direct use of crude oil). For countries outside the OECD, apparent consumption is either a reported figure or is derived from refined product output plus refined product imports minus refined product exports, with stock levels assumed to remain the same. Apparent consumption also includes, where available, liquefied petroleum gases sold directly from natural gas processing plants for fuel or chemical uses.

- **appliance**

A piece of equipment, commonly powered by electricity, used to perform a particular energy-driven function. Examples of common appliances are refrigerators, clothes washers and dishwashers, conventional ranges/ovens and microwave ovens, humidifiers and dehumidifiers, toasters, radios, and televisions. Thus, equipment such as central heating and air conditioning systems and water heaters, which are connected to distribution systems inherent to their purposes, are not considered appliances.

- **appliance efficiency index**

A relative comparison of trends in new-model efficiencies for major appliances and energy-using equipment. The base year for relative comparisons was 1972 (1972=100). Efficiencies for each year were efficiencies of different model types that were weighted by their market shares.

- **appliance efficiency standards**

The National Appliance Energy Conservation Act of 1987 established minimum efficiency standards for major home appliances, including furnaces, central and room air conditioners, refrigerators, freezers, water heaters, dishwashers, and heat pumps. Most of the standards took effect in 1990. The standards for clothes washers, dishwashers, and ranges took effect in 1988, because they required only minor changes in product design, such as eliminating pilot lights and requiring cold water rinse options. The standards for central air conditioners and furnaces took effect in 1992, because it took longer to redesign these products. Appliance efficiency standards for refrigerators took effect in 1993.

- **aromatics**

Hydrocarbons characterized by unsaturated ring structures of carbon atoms. Commercial petroleum aromatics are benzene, toluene and xylene (BTX).

- **as received coal**
Coal in the condition as received by the user.

- **asbestos**
A group of naturally occurring minerals that separate into long, thin fibers. Asbestos was used for many years to insulate and fireproof buildings. In the 1989 CBECS, information on asbestos in buildings was collected for the U.S. Environmental Protection Agency (EPA) Asbestos treatment methods include removal, encapsulation or sealing, and enclosure behind a permanent barrier.

- **ash**
Impurities consisting of silica, iron, alumina, and other noncombustible matter that are contained in coal. Ash increases the weight of coal, adds to the cost of handling, and can affect its burning characteristics Ash content is measured as a percent by weight of coal on an "as received" or a "dry" (moisture-free, usually part of a laboratory analysis) basis.

- **asphalt**
A dark brown-to-black cement-like material obtained by petroleum processing and containing bitumens as the predominant component; used primarily for road construction. It includes crude asphalt as well as the following finished products: cements, fluxes, the asphalt content of emulsions (exclusive of water), and petroleum distillates blended with asphalt to make cutback asphalts.

- **assembly identifier**
A unique string of alphanumeric characters that identifies an assembly, bundle, or canister for a specific reactor in which it has been irradiated.

- **assembly type**
Each assembly is characterized by a fabricator, rod-array size, and model type. An eight-digit assembly type code is assigned to each assembly type based on certain distinguishing characteristics, such as the number of rods per assembly, fuel rod diameter, cladding type, materials used in fabrication, and other design features.

- **assessment work**
The annual or biennial work performed on a mining claim (or claims), after claim location and before patent, to benefit or develop the claim and to protect it from relocation by third parties.

- **assistance for heating in winter**
Assistance from the Low-Income Home Energy Assistance Program (LIHEAP). The purpose of LIHEAP

is to assist eligible households to meet the costs of home energy, i.e., a source of heating or cooling residential buildings.

- **assistance for weatherization of residence**

The household received services free, or at a reduced cost, from the Federal, State, or local Government. Any of the following services could have been received:

- Insulation in the attic, outside wall, or basement/crawlspace below the floor of the house
- Insulation around the hot water heater
- Repair of broken windows or doors to keep out the cold or hot weather
- Weather stripping or caulking around any windows or doors to the outside
- Storm doors or windows added
- Repair of broken furnace Furnace tune up and/or modifications
- Other home energy-saving devices.

- **associated-dissolved natural gas**

Natural gas that occurs in crude oil reservoirs either as free gas (associated) or as gas in solution with crude oil (dissolved gas).

- **astm**

The acronym for the American Society for Testing and Materials.

- **at wt**

The abbreviation for atomic weight.

- **atmospheric crude oil distillation**

The refining process of separating crude oil components at atmospheric pressure by heating to temperatures of about 600 degrees to 750 degrees Fahrenheit (depending on the nature of the crude oil and desired products) and subsequent condensing of the fractions by cooling.

- **auger mine**

A surface mine in which the coal bed is removed by means of a large diameter drill. Usually operated only when the overburden becomes too thick for economical strip mining.

- **authorized cash distribution to municipality**

The authorized cash distributions to the municipality from the earned surplus of the utility department.

- **automatic set-back or clock thermostat**

A thermostat that can be set to turn the heating/cooling system off and on at certain predetermined times.

- **automobile and truck classifications**

Vehicle classifications for automobiles and light duty trucks were obtained from the EPA (Environmental Protection Agency) mileage guide book. Almost every year there are small changes in the classifications, therefore the categories will change accordingly. The EPA mileage guide can be found at any new car dealership.

- **auxiliary generator**
  A generator at the electric plant site that provides power for the operation of the electrical generating equipment itself, including related demands such as plant lighting, during periods when the electric plant is not operating and power is unavailable from the grid. A black start generator used to start main central station generators is considered to be an auxiliary generator.

- **available but not needed capability**
  Net capability of main generating units that are operable but not considered necessary to carry load and cannot be connected to load within 30 minutes.

- **average daily production**
  The ratio of the total production at a mining operation to the total number of production days worked at the operation.

- **average delivered price**
  The weighted average of all contract price commitments and market price settlements in a delivery year.

- **average household energy expenditures**
  A ratio estimate defined as the total household energy expenditures divided by the total number of households.

- **average mine price**
  The ratio of the total value of the coal produced at the mine to the total production tonnage.

- **average production per miner per day**
  The product of the average production per miner per hour at a mining operation and the average length of a production shift at the operation.

- **average production per miner per hour**
  The ratio of the total production at a mining operation to the total direct labor hours worked at the operation.

- **average revenue per kilowatt hour**
  The average revenue per kilowatt hour of electricity sold by sector (residential, commercial, industrial, or other) and geographic area (State, Census division, and national) is calculated by dividing the total monthly revenue by the corresponding total monthly sales for each sector and geographic area.

- **average stream flow**
  The rate, usually expressed in cubic feet per second, at which

water passes a given point in a stream over a set period of time.

- **average vehicle fuel consumption**

A ratio estimate defined as total gallons of fuel consumed by all vehicles divided by: (1) the total number of vehicles (for average fuel consumption per vehicle) or (2) the total number of households (for average fuel consumption per household).

- **average vehicle miles traveled**

A ratio estimate defined as total miles traveled by all vehicles, divided by: (1) the total number of vehicles (for average miles traveled per vehicle) or (2) the total number of households (for average miles traveled per household).

- **average water conditions**

The amount and distribution of precipitation within a drainage basin and the run off conditions present as determined by reviewing the area water supply records over a long period of time.

- **aviation gasoline (finished)**

A complex mixture of relatively volatile hydrocarbons with or without small quantities of additives, blended to form a fuel suitable for use in aviation reciprocating engines. Fuel specifications are provided in ASTM Specification D 910 and Military Specification MIL-G-5572.

- **aviation gasoline blending components**

Naphthas that will be used for blending or compounding into finished aviation gasoline (e.g., straight run gasoline, alkylate, reformate, benzene, toluene, and xylene). Excludes oxygenates (alcohols, ethers), butane, and pentanes plus. Oxygenates are reported as other hydrocarbons, hydrogen, and oxygenates.

- **backup fuel**

In a central heat pump system, the fuel used in the furnace that takes over the space heating when the outdoor temperature drops below that which is feasible to operate a heat pump.

- **backup generator**

A generator that is used only for test purposes, or in the event of an emergency, such as a shortage of power needed to meet customer load requirements.

- **backup power**

Electric energy supplied by a utility to replace power and energy lost during an unscheduled equipment outage.

- **balancing item**

Represents differences between the sum of the components of natural gas supply and the sum of the components of natural gas disposition. These differences may be due to quantities lost or to the effects of data reporting problems. Reporting problems include differences due to the net result of conversions of flow data metered at varying temperature and pressure bases and converted to a standard temperature and pressure base; the effect of variations in company accounting and billing practices; differences between billing cycle and calendar period time frames; and imbalances resulting from the merger of data reporting systems that vary in scope, format, definitions, and type of respondents.

- **barrel**

A unit of volume equal to 42 U.S. gallons.

- **barrels per calendar day**

The amount of input that a distillation facility can process under usual operating conditions. The amount is expressed in terms of capacity during a 24-hour period and reduces the maximum processing capability of all units at the facility under continuous operation to account for the following limitations that may delay, interrupt, or slow down production. The capability of downstream processing units to absorb the output of crude oil processing facilities of a given refinery. No reduction is necessary for intermediate streams that are distributed to other than downstream facilities as part of a refinery's normal operation;

- the types and grades of inputs to be processed;
- the types and grades of products expected to be manufactured;
- the environmental constraints associated with refinery operations;
- the reduction of capacity for scheduled downtime due to such conditions as routine inspection, maintenance, repairs, and turnaround; and
- the reduction of capacity for unscheduled downtime due to such conditions as mechanical problems, repairs, and slowdowns.

- **barrels per stream day**

The maximum number of barrels of input that a distillation facility can process within a 24-hour period when running at full capacity under optimal crude and product slate conditions with no allowance for downtime.

- **base (cushion) gas**

The volume of gas needed as a permanent inventory to maintain adequate reservoir pressures and deliverability rates throughout the withdrawal season. All native gas is included in the base gas volume.

base bill
A charge calculated by taking the rate from the appropriate electric rate schedule and applying it to the level of consumption.

base load capacity
The generating equipment normally operated to serve loads on an around-the-clock basis.

base load plant
A plant, usually housing high-efficiency steam-electric units, which is normally operated to take all or part of the minimum load of a system, and which consequently produces electricity at an essentially constant rate and runs continuously. These units are operated to maximize system mechanical and thermal efficiency and minimize system operating costs.

base load
The minimum amount of electric power delivered or required over a given period of time at a steady rate.

base period
The period of time for which data used as the base of an index number or other ratio, have been collected. This period is frequently one of a year but it may be as short as one day or as long as the average of a group of years. The length of the base period is governed by the nature of the material under review, the purpose for which the index number (or ratio) is being compiled, and the desire to use a period as free as possible from abnormal influences in order to avoid bias.

base rate
A fixed kilowatt hour charge for electricity consumed that is independent of other charges and/or adjustments.

baseboard heater
As a type of heating equipment, a system in which either electric resistance coils or finned tubes carrying steam or hot water are mounted behind shallow panels along baseboards. Baseboards rely on passive convection to distribute heated air in the space. Electric baseboards are an example of an "Individual Space Heater.

bbl
The abbreviation for barrel(s).

bbl/d
The abbreviation for barrel(s) per day.

bbl/sd
The abbreviation for barrel(s) per stream day

bcf
The abbreviation for billion cubic feet.

benzene (c6h6)
An aromatic hydrocarbon present in small proportion in some crude

oils and made commercially from petroleum by the catalytic reforming of naphthenes in petroleum naphtha. Also made from coal in the manufacture of coke. Used as a solvent in the manufacture of detergents, synthetic fibers, petrochemicals, and as a component of high-octane gasoline.

- **bi-fuel vehicle**

A motor vehicle that operates on two different fuels, but not on a mixture of the fuels. Each fuel is stored in a separate tank.

- **bilateral agreement**

A written statement signed by two parties that specifies the terms for exchanging energy.

- **bilateral electricity contract**

A direct contract between an electric power producer and either a user or broker outside of a centralized power pool or power exchange.

- **billing period**

The time between meter readings. It does not refer to the time when the bill was sent or when the payment was to have been received. In some cases, the billing period is the same as the billing cycle that corresponds closely (within several days) to meter-reading dates. For fuel oil and LPG, the billing period is the number of days between fuel deliveries.

- **biodiesel**

Any liquid biofuel suitable as a diesel fuel substitute or diesel fuel additive or extender. Biodiesel fuels are typically made from oils such as soybeans, rapeseed, or sunflowers, or from animal tallow. Biodiesel can also be made from hydrocarbons derived from agricultural products such as rice hulls.

- **biofuels**

Liquid fuels and blending components produced from biomass (plant) feed stocks, used primarily for transportation.

- **biomass energy**

Biomass is matter usually thought of as garbage. Some of it is just stuff lying around — dead trees, tree branches, yard clippings, left-over crops, wood chips (like in the picture to the right), and bark and sawdust from lumber mills. It can even include used tires and livestock manure.

Your trash, paper products that can't be recycled into other paper products, and other household waste are normally sent to the dump. Your trash contains some types of biomass that can be reused. Recycling biomass for fuel and other uses cuts down on the need for "landfills" to hold garbage.

- **biomass gas**

A medium Btu gas containing methane and carbon dioxide, resulting from the action of microorganisms on organic materials such as a landfill.

- **biomass**

Organic nonfossil material of biological origin constituting a renewable energy source.

- **biomass power plant**

A similar thing can be done at animal feed lots. In places where lots of animals are raised, the animals - like cattle, cows and even chickens - produce manure. When manure decomposes, it also gives off methane gas similar to garbage. This gas can be burned right at the farm to make energy to run the farm.

Using biomass does not add to global warming. Plants use and store carbon dioxide (CO_2) when they grow. This is then released when the plant material is burned. Other plants then use that released CO_2 in growing. So using biomass closes this cycle of storing carbon dioxide. Carbon dioxide is a gas that, when there's too much, can contribute to the "greenhouse effect" and global warming.

So, the use of biomass is environmentally friendly because the biomass is reduced, recycled and then reused. It is also a renewable resource because plants to make biomass can be grown over and over.

Today, new ways of using biomass are still being discovered. One way is to produce ethanol, a liquid alcohol fuel. Ethanol can be used in special types of cars that are made for using alcohol fuel instead of gasoline. The alcohol can also be combined with gasoline. This reduces our dependence on oil - a non-renewable fossil fuel.

- **bitumen**

A naturally occurring viscous mixture, mainly of hydrocarbons heavier than pentane, that may contain sulphur compounds and that, in its natural occurring viscous state, is not recoverable at a commercial rate through a well.

- **bituminous coal**

A dense coal, usually black, sometimes dark brown, often with well-defined bands of bright and dull material, used primarily as fuel in steam-electric power generation, with substantial quantities also used for heat and power applications in manufacturing and to make coke. Bituminous coal is the most abundant coal in active U.S. mining regions. Its moisture content usually is less than 20 percent. The heat content of bituminous coal ranges from 21 to 30 million Btu per ton on a moist, mineral-matter-free basis. The heat content of bituminous coal consumed in the United States averages 24 million Btu per ton, on the as-received basis (i.e., containing both inherent moisture and mineral matter).

- **black liquor**

A byproduct of the paper production process, alkaline spent liquor, that can be used as a source of energy. Alkaline spent liquor is removed from the digesters in the process of chemically pulping wood. After evaporation, the residual "black" liquor is burned as a fuel in a recovery furnace that permits the recovery of certain basic chemicals.

- **black lung benefits**

In the content of the coal operation statement of income, this term refers to all payments, including taxes, made by the company attributable to Black Lung.

- **blast furnace**

A furnace in which solid fuel (coke) is burned with an air blast to smelt ore.

- **blast-furnace gas**

The waste combustible gas generated in a blast furnace when iron ore is being reduced with coke to metallic iron. It is commonly used as a fuel within steel works.

- **blending plant**

A facility that has no refining capability but is either capable of producing finished motor gasoline through mechanical blending or blends oxygenates with motor gasoline.

- **block-rate structure**

An electric rate schedule with a provision for charging a different unit cost for various increasing blocks of demand for energy. A reduced rate may be charged on succeeding blocks.

- **bls**

Bureau of Labor Statistics within the U.S. Department of Labor.

- **boe**

The abbreviation for barrels of oil equivalent (used internationally).

■ **boiler**

A device for generating steam for power, processing, or heating purposes; or hot water for heating purposes or hot water supply. Heat from an external combustion source is transmitted to a fluid contained within the tubes found in the boiler shell. This fluid is delivered to an end-use at a desired pressure, temperature, and quality.

■ **boiler fuel**

An energy source to produce heat that is transferred to the boiler vessel in order to generate steam or hot water. Fossil fuel is the primary energy source used to produce heat for boilers.

■ **boiling-water reactor (bwr)**

A light-water reactor in which water, used as both coolant and moderator, is allowed to boil in the core. The resulting steam can be used directly to drive a turbine.

■ **bonded petroleum imports**

Petroleum imported and entered into Customs bonded storage. These imports are not included in the import statistics until they are:

(1) withdrawn from storage free of duty for use as fuel for vessels and aircraft engaged in international trade; or

(2) withdrawn from storage with duty paid for domestic use.

■ **bone coal**

Coal with a high ash content; it is dull in appearance, hard, and compact.

■ **book value**

The portion of the carrying value (other than the portion associated with tangible assets) prorated in each accounting period, for financial reporting purposes, to the extracted portion of an economic interest in a wasting natural resource.

■ **booked costs**

Costs allocated or assigned to inter-departmental or intra company transactions, such as on-system or synthetic natural gas (SNG) production and company-owned gas used in gas operations and recorded in company books or records for accounting and/or regulatory purposes.

■ **borderline customer**

A customer located in the service area of one utility, but supplied by a neighboring utility through an arrangement between the utilities.

■ **bottled gas, lpg, or propane**

Any fuel gas supplied to a building in liquid form, such as liquefied petroleum gas, propane, or butane. It is usually delivered by tank truck and stored near the building in a tank or cylinder until used.

■ **bottom ash**

Residue mainly from the coal burning process that falls to the bottom of the boiler for removal and disposal.

▪ **bottom-hole contribution**

A payment (either in cash or in acreage) that is required by agreement when a test well is drilled to a specified depth regardless of the outcome of the well and that is made in exchange for well and evaluation data.

▪ **bottoming cycle**

A waste-heat recovery boiler recaptures the unused energy and uses it to produce steam to drive a steam turbine generator to produce electricity.

▪ **bp**

The abbreviation for boiling point.

▪ **branded product**

A refined petroleum product sold by a refiner with the understanding that the purchaser has the right to resell the product under a trademark, trade name, service mark, or other identifying symbol or names owned by such refiner.

▪ **break-even cutoff grade**

The lowest grade of material that can be mined and processed considering all applicable costs, without incurring a loss or gaining a profit.

▪ **breccia**

A coarse-grained clastic rock, composed of angular broken rock fragments held together by a mineral cement or in a fine-grained matrix.

▪ **breeder reactor**

A reactor that both produces and consumes fissionable fuel, especially one that creates more fuel than it consumes. The new fissionable material is created by a process known as breeding, in which neutrons from fission are captured in fertile materials.

▪ **breeze**

The fine screenings from crushed coke. Usually breeze will pass through a 1/2-inch or 3/4-inch screen opening. It is most often used as a fuel source in the process of agglomerating iron ore.

▪ **british thermal unit**

The quantity of heat required to raise the temperature of 1 pound of liquid water by 1 degree Fahrenheit at the temperature at which water has its greatest density (approximately 39 degrees Fahrenheit).

▪ **btu conversion factors**

Btu conversion factors for site energy are as follows:

electricity	3,412 btu/ kilowatthour
natural gas	1,031 btu/cubic foot
fuel oil no.1	135,000 btu/gallon
kerosene	135,000 btu/gallon
fuel oil no.2	138,690 btu/gallon
lpg (propane)	91,330 btu/gallon
wood	20 million btu/ cord

■ **btu per cubic foot**

The total heating value, expressed in Btu, produced by the combustion, at constant pressure, of the amount of the gas that would occupy a volume of 1 cubic foot at a temperature of 60 degrees F if saturated with water vapor and under a pressure equivalent to that of 30 inches of mercury at 32 degrees F and under standard gravitational force (980.665 cm. per sec. squared) with air of the same temperature and pressure as the gas, when the products of combustion are cooled to the initial temperature of gas and air when the water formed by combustion is condensed to the liquid state. (Sometimes called gross heating value or total heating value.)

■ **btu**

The abbreviation for British thermal unit(s).

■ **btx**

The acronym for the commercial petroleum aromatics—benzene, toluene, and xylene.

■ **budget plan**

An agreement between the household and the utility company or fuel supplier that allows the household to pay the same amount for fuel for each month for a number of months.

■ **building shell (envelope) dsm program**

A DSM program that promotes reduction of energy consumption through improvements to the building envelope. Includes installations of insulation, weather stripping, caulking, window film, and window replacement.

■ **building shell conservation feature**

A building feature designed to reduce energy loss or gain through the shell or envelope of the building. Data collected by EIA on the following specific building shell energy conservation features: roof, ceiling, or wall insulation; storm windows or double- or triple-paned glass (multiple glazing); tinted or reflective glass or shading films; exterior or interior shadings or awnings; and weather stripping or caulking.

■ **built-in electric units**

An individual-resistance electric-heating unit that is permanently installed in the floors, walls, ceilings, or baseboards and is part of the electrical installation of the building. Electric-heating devices that are plugged into an electric socket or outlet are not considered built in.

■ **bulk power transactions**

The wholesale sale, purchase, and interchange of electricity among electric utilities. Bulk power transactions are used by electric utilities for many different aspects of electric utility operations, from maintaining load to reducing costs.

- **bulk sales**

Wholesale sales of gasoline in individual transactions which exceed the size of a truckload.

- **bulk station**

A facility used primarily for the storage and/or marketing of petroleum products, which has a total bulk storage capacity of less than 50,000 barrels and receives its petroleum products by tank car or truck.

- **bulk terminal**

A facility used primarily for the storage and/or marketing of petroleum products, which has a total bulk storage capacity of 50,000 barrels or more and/or receives petroleum products by tanker, barge, or pipeline.

- **bundled utility service (electric)**

A means of operation whereby energy, transmission, and distribution services, as well as ancillary and retail services, are provided by one entity.

- **bunker fuels**

Fuel supplied to ships and aircraft, both domestic and foreign, consisting primarily of residual and distillate fuel oil for ships and kerosene-based jet fuel for aircraft. The term "international bunker fuels" is used to denote the consumption of fuel for international transport activities. Note: For the purposes of greenhouse gas emissions inventories, data on emissions from combustion of international bunker fuels are subtracted from national emissions totals. Historically, bunker fuels have meant only ship fuel.

- **burn days**

The number of days the station could continue to operate by burning coal already on hand assuming no additional deliveries of coal and an average consumption rate.

- **burnup**

Amount of thermal energy generated per unit mass of fuel, expressed as Gigawatt-Days Thermal per Metric Ton of Initial Heavy Metal (GWDT/MTIHM), rounded to the nearest gigawatt day.

- **bus**

An electrical conductor that serves as a common connection for two or more electrical circuits.

- **butane (c4h10)**

A normally gaseous straight-chain or branch-chain hydrocarbon extracted from hydrocarbon extracted from natural gas or refinery gas streams. It includes isobutane and normal butane and is designated in ASTM Specification D1835 and Gas Processors Association Specifications for commercial butane.

▪ **butylene (c4h8)**

An olefinic hydrocarbon recovered from refinery processes.

▪ **buy-back oil**

Crude oil acquired from a host government whereby a portion of the government's ownership interest in the crude oil produced in that country may or should be purchased by the producing firm.

▪ **bypassed footage**

Bypassed footage is the footage in that section of hole that is abandoned as the result of remedial sidetrack drilling operations.

▪ **byproduct**

A secondary or additional product resulting from the feedstock use of energy or the processing of nonenergy materials. For example, the more common byproducts of coke ovens are coal gas, tar, and a mixture of benzene, toluene, and xylenes (BTX).

▪ **c4h**

A mixture of light hydrocarbons that have the general formula C_4H_n, where **n** is the number of hydrogen atoms per molecule. Examples include butane (C_4H_{10}) and butylene (C_4H_8).

▪ **calcination**

A process in which a material is heated to a high temperature without fusing, so that hydrates, carbonates, or other compounds are decomposed and the volatile material is expelled.

▪ **calcium sulfate**

A white crystalline salt, insoluble in water. Used in Keene's cement, in pigments, as a paper filler, and as a drying agent.

▪ **calcium sulfite**

A white powder, soluble in diluted sulfuric acid. Used in the sulfite process for the manufacture of wood pulp.

▪ **california power exchange**

A State-chartered, non-profit corporation which provides day-ahead and hour-ahead markets for energy and ancillary services in accordance with the power exchange tariff. The power exchange is a scheduling coordinator and is independent of both the independent system operator and all other market participants.

▪ **canadian deuterium uranium reactor (candu)**

Uses heavy water or deuterium oxide (D_2O), rather than light water (H_2O), as the coolant and moderator. Deuterium is an isotope of hydrogen that has a different neutron absorption spectrum from that of ordinary hydrogen. In a deuterium-moderated-reactor, fuel made from natural uranium (0.71 U-235) can sustain a chain reaction.

- **cannel coal**

  A compact, tough variety of coal, originating from organic spore residues, that is non- caking, contains a high percentage of volatile matter, ignites easily, and burns with a luminous smoky flame.

- **capable of being fueled**

  A vehicle is capable of being fueled by a particular fuel(s) if that vehicle has the engine components in place to make operation possible on the fuel(s). The vehicle does not necessarily have to run on the fuel(s) in order for that vehicle to be considered capable of being fueled by the fuel(s). For example, a vehicle that is equipped to operate on either gasoline or natural gas but normally operates on gasoline is considered to be capable of being fueled by gasoline and natural gas.

- **capacity (purchased)**

  The amount of energy and capacity available for purchase from outside the system.

- **capacity charge**

  An element in a two-part pricing method used in capacity transactions (energy charge is the other element). The capacity charge, sometimes called Demand Charge, is assessed on the amount of capacity being purchased.

- **capacity factor**

  The ratio of the electrical energy produced by a generating unit for the period of time considered to the electrical energy that could have been produced at continuous full power operation during the same period.

- **capacity transaction**

  The acquisition of a specified quantity of generating capacity from another utility for a specified period of time. The utility selling the power is obligated to make available to the buyer a specified quantity of power.

- **capacity utilization**

  Capacity utilization is computed by dividing production by productive capacity and multiplying by 100.

- **capital cost**

  The cost of field development and plant construction and the equipment required for industry operations.

- **capital stock**

  Property, plant and equipment used in the production, processing and distribution of energy resources.

- **captive coal**
  Coal produced to satisfy the needs of the mine owner, or of a parent, subsidiary, or other affiliate of the mine owner (for example, steel companies and electricity generators), rather than for open market sale.
- **captive refinery mtbe plants**
  MTBE (methyl tertiary butyl ether) production facilities primarily located within refineries. These integrated refinery units produce MTBE from Fluid Cat Cracker isobutylene with production dedicated to internal gasoline blending requirements.
- **captive refinery oxygenate plants**
  Oxygenate production facilities located within or adjacent to a refinery complex.
- **carbon black**
  An amorphous form of carbon, produced commercially by thermal or oxidative decomposition of hydrocarbons and used principally in rubber goods, pigments, and printer's ink.
- **carbon budget**
  The balance of the exchanges (incomes and losses) of carbon between carbon sinks (e.g., atmosphere and biosphere) in the carbon cycle.
- **carbon cycle**
  All carbon sinks and exchanges of carbon from one sink to another by various chemical, physical, geological, and biological processes.
- **carbon dioxide (co2)**
  A colorless, odorless, non-poisonous gas that is a normal part of Earth's atmosphere. Carbon dioxide is a product of fossil-fuel combustion as well as other processes. It is considered a greenhouse gas as it traps heat (infrared energy) radiated by the Earth into the atmosphere and thereby contributes to the potential for global warming. The global warming potential (GWP) of other greenhouse gases is measured in relation to that of carbon dioxide, which by international scientific convention is assigned a value of one (1).
- **carbon dioxide equivalent**
  The amount of carbon dioxide by weight emitted into the atmosphere that would produce the same estimated radiative forcing as a given weight of another radiatively active gas. Carbon dioxide equivalents are computed by multiplying the weight of the gas being measured (for example, methane) by its estimated global warming potential (which is 21 for methane). "Carbon equivalent units" are defined as carbon dioxide equivalents multiplied by the carbon content of carbon dioxide (i.e., 12/44).

- **carbon intensity**
The amount of carbon by weight emitted per unit of energy consumed. A common measure of carbon intensity is weight of carbon per British thermal unit (Btu) of energy. When there is only one fossil fuel under consideration, the carbon intensity and the emissions coefficient are identical. When there are several fuels, carbon intensity is based on their combined emissions coefficients weighted by their energy consumption levels.

- **carbon output rate**
The amount of carbon by weight per kilowatt hour of electricity produced.

- **carbon sequestration**
The fixation of atmospheric carbon dioxide in a carbon sink through biological or physical processes.

- **carbon sink**
A reservoir that absorbs or takes up released carbon from another part of the carbon cycle. The four sinks, which are regions of the Earth within which carbon behaves in a systematic manner, are the atmosphere, terrestrial biosphere (usually including freshwater systems), oceans, and sediments (including fossil fuels).

- **carburetor**
A fuel delivery device for producing a proper mixture of gasoline vapor and air and for delivering it to the intake manifold of an internal combustion engine. Gasoline is gravity-fed from a reservoir bowl into a throttle bore, where it is allowed to evaporate into the stream of air being inducted by the engine.

- **carrying costs**
Costs incurred in order to retain exploration and property rights after acquisition but before production has occurred. Such costs include legal costs for title defense, ad valorem taxes on non-producing mineral properties, shut-in royalties, and delay rentals.

- **cash and carry**
Kerosene, fuel oil, or bottled gas (tank or propane) purchased with cash, by check, or by credit card and taken home by the purchaser. The purchaser provides the container or pays extra for the container.

- **casinghead gas (or oil well gas)**
Natural gas produced along with crude oil from oil wells. It contains either dissolved or associated gas or both.

- **cast silicon**
Crystalline silicon obtained by pouring pure molten silicon into a vertical mold and adjusting the temperature gradient along the mold volume during cooling to obtain slow, vertically advancing

crystallization of the silicon. The polycrystalline ingot thus formed is composed of large, relatively parallel, interlocking crystals. The cast ingots are sawed into wafers for further fabrication into photovoltaic cells. Cast silicon wafers and ribbon silicon sheets fabricated into cells are usually referred to as polycrystalline photovoltaic cells.

- **catalyst coke**

In many catalytic operations (e.g., catalytic cracking), carbon is deposited on the catalyst, thus deactivating the catalyst. The catalyst is reactivated by burning off the carbon, which is used as a fuel in the refining process. This carbon or coke is not recoverable in a concentrated form.

- **catalytic converter**

A device containing a catalyst for converting automobile exhaust into mostly harmless products.

- **catalytic cracking**

The refining process of breaking down the larger, heavier, and more complex hydrocarbon molecules into simpler and lighter molecules. Catalytic cracking is accomplished by the use of a catalytic agent and is an effective process for increasing the yield of gasoline from crude oil. Catalytic cracking processes fresh feeds and recycled feeds.

- **catalytic hydrocracking**

A refining process that uses hydrogen and catalysts with relatively low temperatures and high pressures for converting middle boiling or residual material to high octane gasoline, reformer charge stock, jet fuel, and /or high grade fuel oil. The process uses one or more catalysts, depending on product output, and can handle high sulfur feed stocks without prior desulfurization.

- **catalytic hydrotreating**

A refining process for treating petroleum fractions from atmospheric or vacuum distillation units (e.g., naphthas, middle distillates, reformer feeds, residual fuel oil, and heavy gas oil) and other petroleum (e.g., cat cracked naphtha, coker naphtha, gas oil, etc.) in the presence of catalysts and substantial quantities of hydrogen. Hydrotreating includes desulfurization, removal of substances (e.g., nitrogen compounds) that deactivate catalysts, conversion of olefins to paraffins to reduce gum formation in gasoline, and other processes to upgrade the quality of the fractions.

- **catalytic reforming**

A refining process using controlled heat and pressure with catalysts to rearrange certain hydrocarbon molecules, thereby converting paraffinic and naphthenic type hydrocarbons (e.g., low octane gasoline boiling range fractions) into petrochemical feedstocks and higher octane stocks suitable for blending into finished gasoline. Catalytic reforming is reported in two categories. They are:

Low Pressure. A processing unit operating at less than 225 pounds per square inch gauge (PSIG) measured at the outlet separator.

High pressure. A processing unit operating at either equal to or greater than 225 pounds per square inch gauge (PSIG) measured at the outlet separator.

cells

Refers to the un-encapsulated semi-conductor components of the module that convert the solar energy to electricity.

cells to oem (non-pv)

Cells shipped to non-photovoltaic original equipment manufacturers such as boat manufacturers, car manufacturers, etc.

census division

Any of nine geographic areas of the United States as defined by the U.S. Department of Commerce, Bureau of the Census. The divisions, each consisting of several States, are defined as follows:

- **New England:** Connecticut, Maine, Massachusetts, New Hampshire, Rhode Island, and Vermont;
- **Middle Atlantic:** New Jersey, New York, and Pennsylvania;
- **East North Central:** Illinois, Indiana, Michigan, Ohio, and Wisconsin;
- **West North Central:** Iowa, Kansas, Minnesota, Missouri, Nebraska, North Dakota, and South Dakota;
- **South Atlantic:** Delaware, District of Columbia, Florida, Georgia, Maryland, North Carolina, South Carolina, Virginia, and West Virginia;
- **East South Central:** Alabama, Kentucky, Mississippi, and Tennessee;
- **West South Central:** Arkansas, Louisiana, Oklahoma, and Texas;
- **Mountain:** Arizona, Colorado, Idaho, Montana, Nevada, New Mexico, Utah, and Wyoming;
- **Pacific:** Alaska, California, Hawaii, Oregon, and Washington.

census region

Any of four geographic areas of the United States as defined by the U.S. Department of Commerce, Bureau of the Census. The Regions, each consisting of various States selected according to population size and physical location, are defined as follows:

- **Northeast:** Connecticut, Maine, Massachusetts, New

Hampshire, New Jersey, New York, Pennsylvania, Rhode Island, and Vermont.

- **South**: Alabama, Arkansas, Delaware, District of Columbia, Florida, Georgia, Kentucky, Louisiana, Maryland, Mississippi, North Carolina, Oklahoma, South Carolina, Tennessee, Texas, Virginia, and West Virginia.
- **Midwest**: Illinois, Indiana, Iowa, Kansas, Michigan, Minnesota, Missouri, Nebraska, North Dakota, Ohio, South Dakota, and Wisconsin.
- **West**: Alaska, Arizona, California, Colorado, Hawaii, Idaho, Montana, Nevada, New Mexico, Oregon, Utah, Washington, and Wyoming.

■ **census region/division**

An hierarchical organization of the **United States** according to geographic areas and sub-areas as follows:

northeast region

new england division
middle atlantic division

south region

south atlantic division
east south central division
west south central division

midwest region

east north central division
west north central division

west region

mountain division
pacific division

■ **central chiller**

Any centrally located air conditioning system that produces chilled water in order to cool air. The chilled water or cold air is then distributed throughout the building, using pipes or air ducts or both. These systems are also commonly known as "chillers," "centrifugal chillers," "reciprocating chillers," or "absorption chillers." Chillers are generally located in or just outside the building they serve. Buildings receiving district chilled water are served by chillers located at central physical plants.

■ **central cooling**

Cooling of an entire building with a refrigeration unit to condition the air. Typically central chillers and ductwork are present in the centrally cooled building.

■ **central physical plant**

A plant owned by, and on the grounds of, a multi-building facility that provides district heating, district cooling, or electricity to other buildings on the same facility. To qualify as a central plant it must provide district heat, district chilled water, or electricity to at least one other building. The central physical plant may be by itself in a separate building or may be located in a building where other activities occur.

■ **central warm air furnace**

A type of space heating equipment where a central

combustor or resistance unit generally using gas, fuel oil, or electricity provides warm air through ducts leading to the various rooms. Heat pumps are not included in this category. A forced air furnace is one in which a fan is used to force the air through the ducts. In a gravity furnace, air is circulated by gravity, relying on the natural flow of warm air up and cold air down; the warm air rises through ducts and the cold air falls through ducts that return it to the furnace to be reheated and this completes the circulation cycle.

- **centralized water heating system**
  Equipment, to heat and store water for other than space heating purposes, which provides hot water from a single location for distribution throughout a building. A residential type tank water heater is a good example of a centralized water heater.

- **certificate**
  A type of permit for public convenience and necessity issued by a utility commission, which authorizes a utility or regulated company to engage in business, construct facilities, provide some services, or abandon service.

- **certificate requirement**
  The maximum annual volume allowed for sales to resale or direct sale customers under certificate authorizations by the Federal Energy Regulatory Commission.

- **cesspool**
  An underground reservoir for liquid waste, typically household sewage.

- **cfs**
  Cubic feet per second.

- **chained dollars**
  A measure used to express real prices. Real prices are those that have been adjusted to remove the effect of changes in the purchasing power of the dollar; they usually reflect buying power relative to a reference year. Prior to 1996, real prices were expressed in constant dollars, a measure based on the weights of goods and services in a single year, usually a recent year. In 1996, the U.S. Department of Commerce introduced the chained-dollar measure. The new measure is based on the average weights of goods and services in successive pairs of years. It is "chained" because the second year in each pair, with its weights, becomes the first year of the next pair. The advantage of using the chained-dollar measure is that it is more closely related to any given period covered and is therefore subject to less distortion over time.

- **characterization**
  Sampling, monitoring, and analysis activities to determine

the extent and nature of contamination at a facility or site. Characterization provides the necessary technical information to develop, screen, analyze, and select appropriate cleanup techniques.

- **charge capacity**

The input (feed) capacity of the refinery processing facilities.

- **chemical separation**

A process for extracting uranium and plutonium from dissolved spent nuclear fuel and irradiated targets. The fission products that are left behind are high-level waste. Chemical separation is also known as reprocessing.

- **chlorofluorocarbon (cfc)**

Any of various compounds consisting of carbon, hydrogen, chlorine, and flourine used as refrigerants. CFCs are now thought to be harmful to the earth's atmosphere.

- **christmas tree**

The valves and fittings installed at the top of a gas or oil well to control and direct the flow of well fluids.

- **cif (cargo, insurance and freight)**

CIF refers to cargos for which the seller pays for the transportation and insurance up to the port of destination.

- **cif (cost, insurance, freight)**

This term refers to a type of sale in which the buyer of the product agrees to pay a unit price that includes the f.o.b. value of the product at the point of origin plus all costs of insurance and transportation. This type of a transaction differs from a "delivered" purchase, in that the buyer accepts the quantity as determined at the loading port (as certified by the Bill of Lading and Quality Report) rather than pay based on the quantity and quality ascertained at the unloading port. It is similar to the terms of an f.o.b. sale, except that the seller, as a service for which he is compensated, arranges for transportation and insurance.

- **circuit**

A conductor or a system of conductors through which electric current flows.

- **circuit-mile**

The total length in miles of separate circuits regardless of the number of conductors used per circuit.

- **citygate**

A point or measuring station at which a distributing gas utility receives gas from a natural gas pipeline company or transmission system.

- **class rate schedule**
  An electric rate schedule applicable to one or more specified classes of service, groups of businesses, or customer uses.

- **classes of service**
  Customers grouped by similar characteristics in order to be identified for the purpose of setting a common rate for electric service. Usually classified into groups identified as residential, commercial, industrial, and other.

- **clean development mechanism (cdm)**
  A Kyoto Protocol program that enables industrialized countries to finance emissions-avoiding projects in developing countries and receive credit for reductions achieved against their own emissions limitation targets.

- **climate change**
  A term used to refer to all forms of climatic inconsistency, but especially to significant change from one prevailing climatic condition to another. In some cases, "climate change" has been used synonymously with the term "global warming"; scientists, however, tend to use the term in a wider sense inclusive of natural changes in climate, including climatic cooling.

- **clinker**
  Powdered cement, produced by heating a properly proportioned mixture of finely ground raw materials (calcium carbonate, silica, alumina, and iron oxide) in a kiln to a temperature of about 2,700 degrees Fahrenheit.

- **cloud condensation nuclei**
  Aerosol particles that provide a platform for the condensation of water vapor, resulting in clouds with higher droplet concentrations and increased albedo.

- **co control period ("seasons")**
  The portion of the year in which a CO non-attainment area is prone to high ambient levels of carbon monoxide. This portion of the year is to be specified by the Environmental Protection Agency but is to be not less than 4 months in length.

- **coal**
  A readily combustible black or brownish-black rock whose composition, including inherent moisture, consists of more than 50 percent by weight and more than 70 percent by volume of carbonaceous material. It is formed from plant remains that have been compacted, hardened, chemically altered, and metamorphosed by heat and pressure over geologic time.

- **coal analysis**
  Determines the composition and properties of coal so it can be ranked and used most effectively:

- **coal bed** A bed or stratum of coal. Also called a coal seam.
- **coal bed degasification** This refers to the removal of methane or coal bed gas from a coal mine before or during mining.
- **coal bed methane** Methane is generated during coal formation and is contained in the coal microstructure. Typical recovery entails pumping water out of the coal to allow the gas to escape. Methane is the principal component of natural gas. Coal bed methane can be added to natural gas pipelines without any special treatment.

■ **coal briquets**

Anthracite, bituminous, and lignite briquets comprise the secondary solid fuels manufactured from coal by a process in which the coal is partly dried, warmed to expel excess moisture, and then compressed into briquets, usually without the use of a binding substance. In the reduction of briquets to coal equivalent, different conversion factors are applied according to their origin from hard coa!, peat, brown coal, or lignite.

■ **coal carbonized**

The amount of coal decomposed into solid coke and gaseous products by heating in a coke oven in a limited air supply or in the absence of air.

■ **coal chemicals**

Coal chemicals are obtained from the gases and vapor recovered from the manufacturing of coke. Generally, crude tar, ammonia, crude light oil, and gas are the basic products recovered. They are refined or processed to yield a variety of chemical materials.

■ **coal consumption**

The quantity of coal burned for the generation of electric power (in short tons), including fuel used for maintenance of standby service.

■ **coal delivered**

Coal which has been delivered from the coal supplier to any site belonging to the electric power company.

■ **coal face**

This is the exposed area from which coal is extracted.

■ **coal financial reporting regions**

A geographic classification of areas with coal resources which is used for financial reporting of coal statistics.

- **Eastern Region.** Consists of the Appalachian Coal Basin. The following comprise the Eastern Region: Alabama, eastern Kentucky, Georgia, Maryland, Mississippi, Ohio, Pennsylvania, Virginia, Tennessee, North Carolina, and West Virginia.

- **Midwest Region.** Consists of the Illinois and Michigan Coal Basins. The following comprise the Midwest Region: Illinois, Indiana, Michigan, and western Kentucky.
- **Western Region.** Consists of the Northern Rocky, Southern Rocky, West Coast Coal Basins and Western Interior. The following comprise the Western Region: Alaska, Arizona, Arkansas, California, Colorado, Idaho, Iowa, Kansas, Louisiana, Missouri, Montana, New Mexico, North Dakota, Oklahoma, Oregon, Texas, South Dakota, Utah, Washington, and Wyoming.

■ **coal fines**

Coal with a maximum particle size usually less than one-sixteenth inch and rarely above one- eighth inch.

■ **coal gas**

Substitute natural gas produced synthetically by the chemical reduction of coal at a coal gasification facility.

■ **coal gasification**

The process of converting coal into gas. The basic process involves crushing coal to a powder, which is then heated in the presence of steam and oxygen to produce a gas. The gas is then refined to reduce sulfur and other impurities. The gas can be used as a fuel or processed further and concentrated into chemical or liquid fuel.

■ **coal grade**

This classification refers to coal quality and use.

- **Briquettes** are made from compressed coal dust, with or without a binding agent such as asphalt.
- **Cleaned coal or prepared coal** has been processed to reduce the amount of impurities present and improve the burning characteristics.
- **Compliance coal** is a coal, or a blend of coal, that meets sulfur dioxide emission standards for air quality without the need for flue-gas desulfurization.
- **Culm and silt** are waste materials from preparation plants. In the anthracite region, culm consists of coarse rock fragments containing as much as 30 percent small-sized coal. Silt is a mixture of very fine coal particles (approx. 40 percent) and rock dust that has settled out from waste water from the plants. The terms culm and silt are sometimes used interchangeably and are sometimes called refuse. Culm and silt have a heat value ranging from 8 to 17 million Btu per ton.
- **Low-sulfur coal** generally contains 1 percent or less

sulfur by weight. For air quality standards, "low sulfur coal" contains 0.6 pounds or less sulfur per million Btu, which is equivalent to 1.2 pounds of sulfur dioxide per million Btu.

- **Metallurgical coal (or coking coal)** meets the requirements for making coke. It must have a low ash and sulfur content and form a coke that is capable of supporting the charge of iron ore and limestone in a blast furnace. A blend of two or more bituminous coals is usually required to make coke.
- **Pulverized coal** is a coal that has been crushed to a fine dust in a grinding mill. It is blown into the combustion zone of a furnace and burns very rapidly and efficiently.
- **Slack coal** usually refers to bituminous coal one-half inch or smaller in size.
- **Steam coal** refers to coal used in boilers to generate steam to produce electricity or for other purposes.
- **Stoker coal** refers to coal that has been crushed to specific sizes (but not powdered) for burning on a grate in automatic firing equipment.

▪ coal liquefaction

A chemical process that converts coal into clean-burning liquid hydrocarbons, such as synthetic crude oil and methanol.

▪ coal mining productivity

Coal mining productivity is calculated by dividing total coal production by the total direct labor hours worked by all mine employees.

▪ coal preparation

The process of sizing and cleaning coal to meet market specifications by removing impurities such as rock, sulfur, etc. It may include crushing, screening, or mechanical cleaning.

▪ coal producing districts

A classification of coal fields defined in the Bituminous Coal Act of 1937. The districts were originally established to aid in formulating minimum prices of bituminous and subbituminous coal and lignite. Because much statistical information was compiled in terms of these districts, their use for statistical purposes has continued since the abandonment of that legislation in 1943. District 24 was added for the anthracite-producing district in Pennslyvania.

▪ coal production

The sum of sales, mine consumption, issues to miners, and issues to coke, briquetting, and other ancillary plants at mines. Production data include quantities extracted from surface and underground mines, and normally exclude wastes removed at mines or associated reparation plants.

- **coal rank**

The classification of coals according to their degree of progressive alteration from lignite to anthracite. In the United States, the standard ranks of coal include lignite, subbituminous coal, bituminous coal, and anthracite and are based on fixed carbon, volatile matter, heating value, and agglomerating (or caking) properties.

- **coal sampling**

The collection and proper storage and handling of a relatively small quantity of coal for laboratory analysis. Sampling may be done for a wide range of purposes, such as: coal resource exploration and assessment, characterization of the reserves or production of a mine, to characterize the results of coal cleaning processes, to monitor coal shipments or receipts for adherence to coal quality contract specifications, or to subject a coal to specific combustion or reactivity tests related to the customer's intended use. During pre-development phases, such as exploration and resource assessment, sampling typically is from natural outcrops, test pits, old or existing mines in the region, drill cuttings, or drilled cores. Characterization of a mine's reserves or production may use sample collection in the mine, representative cuts from coal conveyors or from handling and loading equipment, or directly from stockpiles or shipments (coal rail cars or barges). Contract specifications rely on sampling from the production flow at the mining or coal handling facility or at the load out, or from the incoming shipments at the receiver's facility. In all cases, the value of a sample taken depends on its being representative of the coal under consideration, which in turn requires that appropriate sampling procedures be carefully followed. For coal resource and estimated reserve characterization, appropriate types of samples include:

Face channel or channel sample: a sample taken at the exposed coal in a mine by cutting away any loose or weathered coal then collecting on a clean surface a sample of the coal seam by chopping out a channel of uniform width and depth; a face channel or face sample is taken at or near the working face, the most freshly exposed coal where actual removal and loading of mined coal is taking place. any partings greater than 3/8 inch and/or mineral concretions greater than 1/2 inch thick and 2 inches in maximum diameter are normally discarded from a channel sample so as better to represent coal that has been mined, crushed, and screened to remove at least gross non-coal materials.

column sample: a channel or drill core sample taken to represent the entire geologic coal bed; it includes all partings and impurities that may exist in the coalbed.

bench sample: a face or channel sample taken of just that contiguous portion of a coalbed that is considered practical to mine, also known as a "bench;" for example, bench samples may be taken of minable coal where impure coal that makes up part of the geologic coalbed is likely to be left in the mine, or where thick partings split the coal into two or more distinct minable seams, or where extremely thick coalbeds cannot be recovered by normal mining equipment, so that the coal is mined in multiple passes, or benches, usually defined along natural bedding planes.

composite sample: a recombined coalbed sample produced by averaging together thickness-weighted coal analyses from partial samples of the coalbed, such as from one or more bench samples, from one or more mine exposures or outcrops where the entire bed could not be accessed in one sample, or from multiple drill cores that were required to retrieve all local sections of a coal seam.

▪ coal stocks

Coal quantities that are held in storage for future use and disposition. Note: When coal data are collected for a particular reporting period (month, quarter, or year), coal stocks are commonly measured as of the last day of this period.

▪ coal sulfur

Coal sulfur occurs in three forms: organic, sulfate, and pyritic. Organic sulfur is an integral part of the coal matrix and cannot be removed by conventional physical separation. Sulfate sulfur is usually negligible. Pyritic sulfur occurs as the minerals pyrite and marcasite; larger sizes generally can be removed by cleaning the coal.

▪ coal type

The classification is based on physical characteristics or microscopic constituents. Examples of coal types are banded coal, bright coal, cannel coal, and splint coal. The term is also used to classify coal according to heat and sulfur content.

▪ coal zone

A series of laterally extensive and (or) lenticular coal beds and associated strata that arbitrarily can be viewed as a unit. Generally, the coal beds in a coal zone are assigned to the same geologic member or formation.

■ **coal-fired power plant works**

Electricity is produced at a coal-fired fossil plant by the process of heating water in a boiler to produce steam. The steam, under tremendous pressure, flows into a turbine, which spins a generator to produce electricity.

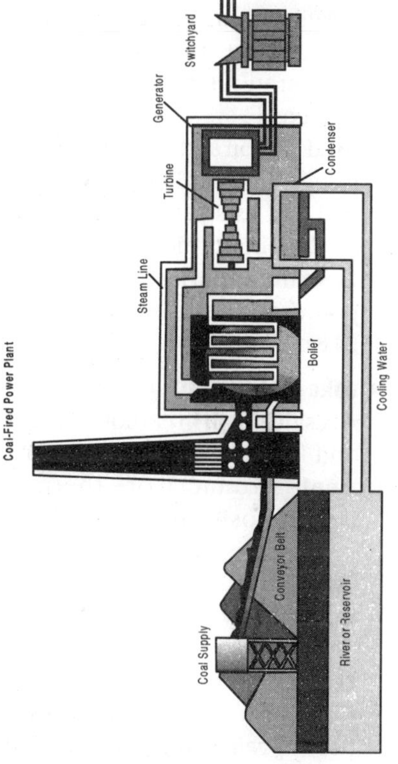

At the Kingston Fossil Plant near Knoxville, Tennessee, for example, each boiler heats the water to about 1,000 degrees Fahrenheit (540 degrees Celsius) to create steam. The steam is sent into the turbines at pressures of more than 1,800 pounds per square inch (130 kilograms per square centimeter). This one plant generates about 10 billion kilowatt-hours a year, or enough electricity to supply 700,000 homes. To meet this demand, Kingston burns about 14,000 tons of coal a day, an amount that would fill 140 railroad cars.

■ **coal-producing regions**

Appalachian Region. Consists of Alabama, Georgia, Eastern Kentucky, Maryland, North Carolina, Ohio, Pennsylvania, Tennessee, Virginia, and West Virginia.

Interior Region (with Gulf Coast). Consists of Arkansas, Illinois, Indiana, Iowa, Kansas, Louisiana, Michigan, Mississippi, Missouri, Oklahoma, Texas, and Western Kentucky.

Western Region. Consists of Alaska, Arizona, Colorado, Montana, New Mexico, North Dakota, Utah, Washington, and Wyoming.

■ **code of federal regulations**

A compilation of the general and permanent rules of the executive departments and agencies of the Federal Government as published in the Federal Register. The code is divided into 50 titles that represent broad areas subject to Federal regulation. Title 18 contains the FERC regulations.

■ **cofiring**

The process of burning natural gas in conjunction with another fuel to reduce air pollutants.

■ **cogeneration system**

A system using a common energy source to produce both electricity and steam for other uses, resulting in increased fuel efficiency.

■ **cogeneration**

The production of electrical energy and another form of useful energy (such as heat or steam) through the sequential use of energy.

■ **cogenerator**

A generating facility that produces electricity and another form of useful thermal energy (such as heat or steam), used for industrial, commercial, heating, or cooling purposes. To receive status as a qualifying facility (QF) under the Public Utility Regulatory Policies Act (PURPA), the facility must produce electric energy and "another form of useful thermal energy through the sequential use of energy" and meet certain ownership, operating, and efficiency criteria established by the Federal Energy Regulatory Commission (FERC).

■ **coincidental demand**

The sum of two or more demands that occur in the same time interval.

■ **coincidental peak load**

The sum of two or more peak loads that occur in the same time interval.

■ **coke (coal)**

A solid carbonaceous residue derived from low-ash, low-sulfur bituminous coal from which the volatile constituents are driven off by baking in an oven at temperatures as high as 2,000 degrees Fahrenheit so that the fixed carbon and residual ash are fused together. Coke is used as a fuel and as a reducing agent in smelting iron ore in a blast furnace. Coke from coal is grey, hard, and porous and has a heating value of 24.8 million Btu per ton.

■ **coke (petroleum)**

A residue high in carbon content and low in hydrogen that is the final product of thermal decomposition in the condensation process in cracking. This product is reported as marketable coke or catalyst coke. The conversion is 5 barrels (of 42 U.S. gallons each) per short ton. Coke from petroleum has a heating value of 6.024 million Btu per barrel.

■ **coke breeze**

The term refers to the fine sizes of coke, usually less than one-half inch, that are recovered from coke plants. It is commonly used for sintering iron ore.

- **coke button**
A button-shaped piece of coke resulting from standard laboratory tests that indicates the coking or free-swelling characteristics of a coal; expressed in numbers and compared with a standard.

- **coke oven gas**
The mixture of permanent gases produced by the carbonization of coal in a coke oven at temperatures in excess of 1,000 degrees Celsius.

- **coke plants**
Plants where coal is carbonized for the manufacture of coke in slot or beehive ovens.

- **coking coal**
Bituminous coal suitable for making coke.

- **coking**
Thermal refining processes used to produce fuel gas, gasoline blend stocks, distillates, and petroleum coke from the heavier products of atmospheric and vacuum distillation. Includes:

- **cold-deck imputation**
A statistical procedure that replaces a missing value of an item with a constant value from an external source such as a value from a previous survey.

- **combined cycle**
An electric generating technology in which electricity is produced from otherwise lost waste heat exiting from one or more gas (combustion) turbines. The exiting heat is routed to a conventional boiler or to a heat recovery steam generator for utilization by a steam turbine in the production of electricity. This process increases the efficiency of the electric generating unit.

- **combined cycle unit**
An electric generating unit that consists of one or more combustion turbines and one or more boilers with a portion of the required energy input to the boiler(s) provided by the exhaust gas of the combustion turbine(s).

- **combined heat and power (chp) plant**
A plant designed to produce both heat and electricity from a single heat source. Note: This term is being used in place of the term "cogenerator" that was used by EIA in the past. CHP better describes the facilities because some of the plants included do not produce heat and power in a sequential fashion and, as a result, do not meet the legal definition of cogeneration specified in the Public Utility Regulatory Policies Act (PURPA)

- **combined household energy expenditures**
The total amount of funds spent for energy consumed in, or

delivered to, a housing unit during a given period of time and for fuel used to operate the motor vehicles that are owned or used on a regular basis by the household. The total dollar amount for energy consumed in a housing unit includes state and local taxes but excludes merchandise repairs or special service charges. Electricity, and natural gas expenditures are for the amount of those energy sources consumed. Fuel oil, kerosene, and LPG expenditures are for the amount of fuel purchased, which may differ from the amount of fuel consumed. The total dollar amount of fuel spent for vehicles is the product of fuel consumption and price.

- **combined hydroelectric plant**

A hydroelectric plant that uses both pumped water and natural stream flow for the production of power.

- **combined pumped-storage plant**

A pumped-storage hydroelectric power plant that uses both pumped water and natural stream flow to produce electricity.

- **combustion chamber**

An enclosed vessel in which chemical oxidation of fuel occurs.

- **combustion**

Chemical oxidation accompanied by the generation of light and heat.

- **commercial building**

A building with more than 50 percent of its floor space used for commercial activities. Commercial buildings include, but are not limited to, stores, offices, schools, churches, gymnasiums, libraries, museums, hospitals, clinics, warehouses, and jails. Government buildings are included except for buildings on military bases or reservations.

- **commercial facility**

An economic unit that is owned or operated by one person or organization and that occupies two or more commercial buildings at a single location. A university and a large hospital complex are examples of a commercial multi-building facility.

- **commercial operation (nuclear)**

The phase of reactor operation that begins when power ascension ends and the operating utility formally declares the nuclear power plant to be available for the regular production of electricity. This declaration is usually related to the satisfactory completion of qualification tests on critical components of the unit.

- **commercial sector**

An energy-consuming sector that consists of service-providing facilities and equipment of: businesses; Federal, State, and local governments; and other private and public organizations, such as religious, social, or fraternal groups. The commercial sector includes institutional living quarters. It also includes sewage treatment facilities. Common uses of energy associated with this sector include space heating, water heating, air conditioning, lighting, refrigeration, cooking, and running a wide variety of other equipment.

- **commingling**
The mixing of one utility's generated supply of electric energy with another utility's generated supply within a transmission system.

- **commissioned agent**
An agent who wholesales or retails a refined petroleum product under a commission arrangement. The agent does not take title to the product or establish the selling price, but receives a percentage of fixed fee for serving as an agent.

- **common equity (book value)**
The retained earnings and common stock earnings plus the balances in common equity reserves and all other common stock accounts. This also includes the capital surplus, the paid-in surplus, the premium on common stocks, except those balances specifically related to preferred or preference stocks; less any common stocks held in the treasury.

- **compact fluorescent bulbs**
These are also known as "screw-in fluorescent replacements for incandescent" or "screw-ins." Compact fluorescent bulbs combine the efficiency of fluorescent lighting with the convenience of a standard incandescent bulb. There are many styles of compact fluorescent, including exit light fixtures and floodlights (lamps containing reflectors). Many screw into a standard light socket, and most produce a similar color of light as a standard incandescent bulb. Compact fluorescent bulbs come with ballasts that are electronic (lightweight, instant, no-flicker starting, and 10 to 15% more efficient) or magnetic (much heavier and slower starting). Other types of compact fluorescent bulbs include adaptive circulation and PL and SL lamps and ballasts. Compact fluorescent bulbs are designed for residential uses; they are also used in table lamps, wall sconces, and hall and ceiling fixtures of hotels, motels, hospitals, and other types of commercial buildings with residential-type applications.

- **company automotive (retail) outlet**
Any retail outlet selling motor fuel under the brand name of a

company reporting in the EIA Financial Reporting System.

■ **company-lessee automotive outlet**
One of three types of company automotive (retail) outlets. This type of outlet is operated by an independent marketer who leases the station and land and has use of tanks, pumps, signs, etc. A lessee dealer typically has a supply agreement with a refiner or a distributor and purchases products at dealer tank wagon prices. The term includes outlets operated by commissioned agents and is limited to those dealers who are supplied directly by a refiner or any affiliate or subsidiary company of a refiner.

■ **company-open automotive outlet**
One of three types of company automotive (retail) outlets. This type of outlet is operated by an independent marketer who owns or leases (from a third party that is not a refiner) the station or land of a retail outlet and has use of tanks, pumps, signs, etc. An open dealer typically has a supply agreement with a refiner or a distributor and purchases products based on either rack or dealer tank wagon prices.

■ **company-operated automotive outlet**
One of three types of company automotive (retail) outlets. This type of outlet is operated by salaried or commissioned personnel paid by the reporting company.

■ **company-operated retail outlet**
Any retail outlet (i.e., service station) which sells motor vehicle fuels and is under the direct control of a **firm** that sets the retail product price and directly collects all or part of the retail margin. The category includes retail outlets operated by (1) salaried employees of the firm and/or its subsidiaries and affiliates, (2) licensed or commissioned agents, and/or (3) personnel services contracted by the firm.

■ **competitive transition charge**
A non-bypassable charge levied on each customer of the distribution utility, including those who are served under contracts with nonutility suppliers, for recovery of the utility's **stranded costs** that develop because of competition.

■ **completion (oil/gas production)**
The term refers to the installation of permanent equipment for the production of oil or gas. If a well is equipped to produce only oil or gas from one zone or reservoir, the definition of a "well" (classified as an oil well or gas well) and the definition of a "completion" are

identical. However, if a well is equipped to produce oil and/or gas separately from more than one reservoir, a "well" is not synonymous with a "completion."

- **completion date (oil/gas production)**

The date on which the installation of permanent equipment has been completed as reported to the appropriate regulatory agency. The date of completion of a dry hole is the date of abandonment as reported to the appropriate agency. The date of completion of a service well is the date on which the well is equipped to perform the service for which it was intended.

- **compliance coal**

A coal or a blend of coals that meets sulfur dioxide emission standards for air quality without the need for flue gas desulfurization.

- **compressed natural gas (cng)**

Natural gas which is comprised primarily of methane, compressed to a pressure at or above 2,400 pounds per square inch and stored in special high-pressure containers. It is used as a fuel for natural gas powered vehicles.

- **compressor station**

Any combination of facilities that supply the energy to move gas in transmission or distribution lines or into storage by increasing the pressure.

- **concentrating solar power or solar thermal power system**

A solar energy conversion system characterized by the optical concentration of solar rays through an arrangement of mirrors to generate a high temperature working fluid. Concentrating solar power (but not Solar thermal power) may also refer to a system that focuses solar rays on a photovoltaic cell to increase conversion efficiency.

- **concentrator**

A reflective or refractive device that focuses incident insolation onto an area smaller than the reflective or refractive surface, resulting in increased insolation at the point of focus.

- **concession**

The operating right to explore for and develop petroleum fields in consideration for a share of production in kind (equity oil).

- **concessionary purchases**

The quantity of crude oil exported during a reporting period, which was acquired from the producing government under terms that arise from the firm's participation in a concession. It includes preferential crude where the reporting firm's access to such crude is derived from a former concessionary relationship.

- **condensate (lease condensate)**

A natural gas liquid recovered from associated and nonassociated gas wells from lease separators or field facilities, reported in barrels of 42 U.S. gallons at atmospheric pressure and 60 degrees Fahrenheit.

- **condenser cooling water**

A source of water external to a boiler's feed system is passed through the steam leaving the turbine in order to cool and condense the steam. This reduces the steam's exit pressure and recaptures its heat, which is then used to preheat fluid entering the boiler, thereby increasing the plant's thermodynamic efficiency.

- **conditional energy intensity**

Total consumption of a particular energy source(s) or fuel(s) divided by the total floor space of buildings that use the energy source(s) or fuel(s); i.e., the ratio of consumption to energy source-specific floor space.

- **conditionally effective rates**

An electric rate schedule that has been put into effect by the FERC subject to refund pending final disposition or refilling.

- **conductor**

Metal wires, cables, and bus-bar used for carrying electric current. Conductors may be solid or stranded, that is, built up by a assembly of smaller solid conductors.

- **conference of the parties (cop)**

The collection of nations that have ratified the Framework Convention on Climate Change (FCCC). The primary role of the COP is to keep implementation of the FCCC under review and make the decisions necessary for its effective implementation.

- **configuration maps**

Geographic information containing transmission line, substation, and terminal information. It shows the normal operating voltages and includes information about other operational and political boundaries.

- **congestion**

A condition that occurs when insufficient transfer capacity is available to implement all of the preferred schedules for electricity transmission simultaneously.

- **connected load**

The sum of the continuous ratings or the capacities for a system, part of a system, or a customer's electric power consuming apparatus.

- **connection**

The physical connection (e.g., transmission lines, transformers, switch gear, etc.) between two electric systems permitting the transfer of electric energy in one or both directions.

- **conservation and other dsm**
  This Demand-Side Management category represents the amount of consumer load reduction at the time of system peak due to utility programs that reduce consumer load during many hours of the year. Examples include utility rebate and shared savings activities for the installation of energy efficient appliances, lighting and electrical machinery, and weatherization materials. In addition, this category includes all other Demand-Side Management activities, such as thermal storage, time-of-use rates, fuel substitution, measurement and evaluation, and any other utility-administered Demand-Side Management activity designed to reduce demand and/or electricity use.

- **conservation feature**
  A feature in the building designed to reduce the usage of energy.

- **conservation program**
  A program in which a utility company furnishes home weatherization services free or at reduced cost or provides free or low cost devices for saving energy, such as energy efficient light bulbs, flow restrictors, weather stripping, and water heater insulation.

- **consolidated metropolitan statistical area (cmsa)**
  An area that meets the requirements of a metropolitan statistical area, has a population of one million or more, and consists of two or more component parts that are recognized as **primary metropolitan statistical areas**.

- **construction**
  An energy-consuming subsector of the **industrial sector** that consists of all facilities and equipment used to perform land preparation and construct, renovate, alter, install, maintain, or repair major infrastructure or individual systems therein. Infrastructure includes buildings; industrial plants; and other major structures, such as tanks, towers, monuments, roadways, tunnels, bridges, dams, pipelines, and transmission lines.

- **construction costs (of the electric power industry)**
  All direct and indirect costs incurred in acquiring and constructing electric utility plant and equipment and proportionate shares of common utility plants. included are the cost of land and improvements, nuclear fuel and spare parts, allowance for funds used during construction, and general overheads capitalized, less the cost of acquiring plant and

equipment previously operated in utility service.

- **construction expenditures (of the electric power industry)**

The gross expenditures for construction costs (including the cost of replacing worn out plants), and electric construction costs, and land held for future use.

- **construction pipeline (of a nuclear reactor)**

The various stages involved in the acquisition of a nuclear reactor by a utility. The events that define these stages are the ordering of a reactor, the licensing process, and the physical construction of the nuclear generating unit. A reactor is said to be "in the pipeline" when the reactor is ordered and "out of the pipeline" when it completes low power testing and begins operation toward full power.

- **construction work in progress (cwip)**

The balance shown on a utility's balance sheet for construction work not yet completed but in process. This balance line item may or may not be included in the rate base.

- **constructive surplus or deficit**

The amounts representing the exchange of services, supplies, etc., between the utility department and the municipality and its other departments without charge or at a reduced charge. Charges to this account include utility and other services, supplies, etc., furnished by the utility department to the municipality or its other departments without charge, or the amount of the reduction, if furnished at a reduced charge. Credits to the account consist of services, supplies, office space, etc., furnished by the municipality to the utility department without charge on the amount of the reduction, if furnished at a reduced charge.

- **consumer (energy)**

Any individually metered dwelling, building, establishment, or location using natural gas, synthetic natural gas, and/or mixtures of natural and supplemental gas for feedstock or as fuel for any purpose other than in oil or gas lease operations; natural gas treating or processing plants; or pipeline, distribution, or storage compressors.

- **consumer charge**

An amount charged periodically to a consumer for such utility costs as billing and meter reading, without regard to demand or energy consumption.

- **consumer price index (cpi)**

These prices are collected in 85 urban areas selected to represent all urban consumers about 80 percent of the total U.S. population. The service stations are selected initially and on a replacement basis, in such a way that they represent the purchasing habits of the CPI population. Service stations in the current sample include those providing all types of service (i.e., full, mini, and self service).

- **consumption per square foot**
The aggregate ratio of total consumption for a particular set of buildings to the total floor space of those buildings.

- **continuous delivery energy sources**
Those energy sources provided continuously to a building.

- **continuous mining**
A form of room pillar mining in which a continuous mining machine extracts and removes coal from the working face in one operation; no blasting is required.

- **contract price**
The delivery price determined when a contract is signed. It can be a fixed price or a base price escalated according to a given formula.

- **contract receipts**
Purchases based on a negotiated agreement that generally covers a period of 1 or more years.

- **contracted gas**
Any gas for which Interstate Pipeline has a contract to purchase from any domestic or foreign source that cannot be identified to a specific field or group. This includes tailgate plant purchases, single meter point purchases, pipeline purchases, natural gas imports, SNG purchases, and LNG purchases.

- **contribution to net income**
The FRS (Financial Reporting System survey) segment equivalent to net income. However, some consolidated items of revenue and expense are not allocated to the segments, and therefore they are not equivalent in a strict sense. The largest item not allocated to the segments is interest expense since this is regarded as a corporate level item for FRS purposes.

- **control**
Including the terms "controlling," "controlled by," and "under common control with," means the possession, direct or indirect, of the power to direct or cause the direction of the management and policies of a person, whether through the ownership of voting shares, by contract, or otherwise.

- **control total**
The number of elements in the population or a subset of the population. The sample weights for the observed elements in a survey are adjusted so that they add up to the control total. The value of a control total is obtained

from an outside source. The control totals are given by the number of households in one of the 12 cells by categorizing households by the four Census regions and by three categories of metropolitan status (Metropolitan Statistical Area central city, Metropolitan Statistical Area outside central city, and non Metropolitan Statistical Area). The control totals are obtained from the Current Population Survey.

- **conventional gasoline**
  Finished motor gasoline not included in the oxygenated or reformulated gasoline categories. Note: This category excludes reformulated gasoline blend stock for oxygenate blending (RBOB) as well as other blend stock.

- **conventional hydroelectric plant**
  A plant in which all of the power is produced from natural stream flow as regulated by available storage.

- **conventional mill (uranium)**
  A facility engineered and built principally for processing of uraniferous ore materials mined from the earth and the recovery, by chemical treatment in the mill's circuits, of uranium and/or other valued coproduct components from the processed one.

- **conventional mining**
  The oldest form of room pillar mining, which consists of a series of operations that involve cutting the coal bed, so it breaks easily when blasted with explosives or high pressure air, and then loading the broken coal.

- **conventional thermal electricity generation**
  Electricity generated by an electric power plant using coal, petroleum, or gas as its source of energy.

- **conventionally fueled vehicle**
  A vehicle that runs on petroleum-based fuels such as motor gasoline or diesel fuel.

- **conversion company**
  An organization that performs vehicle conversions on a commercial basis.

- **conversion factor**
  A number that translates units of one measurement system into corresponding values of another measurement system.

- **converted (alternative-fuel) vehicle**
  A vehicle originally designed to operate on gasoline/diesel that was modified or altered to run on an alternative fuel after its initial delivery to an end-user.

- **cooling**
  Conditioning of room air for human comfort by a refrigeration

unit (such as an air conditioner or heat pump) or by circulating chilled water through a central cooling or district cooling system. Use of fans or blowers by themselves, without chilled air or water, is not included in this definition of cooling.

- **cooling degree-days**

A measure of how warm a location is over a period of time relative to a base temperature, most commonly specified as 65 degrees Fahrenheit. The measure is computed for each day by subtracting the base temperature (65 degrees) from the average of the day's high and low temperatures, with negative values set equal to zero. Each day's cooling degree-days are summed to create a cooling degree-day measure for a specified reference period. Cooling degree-days are used in energy analysis as an indicator of air conditioning energy requirements or use.

- **cooling pond**

A natural or man made body of water that is used for dissipating waste heat from power plants.

- **cooling system**

An equipment system that provides water to the condensers and includes water intakes and outlets; cooling towers; and ponds, pumps, and pipes.

- **cooperative electric utility**

An electric utility legally established to be owned by and operated for the benefit of those using its service. The utility company will generate, transmit, and/or distribute supplies of electric energy to a specified area not being serviced by another utility. Such ventures are generally exempt from Federal income tax laws. Most electric cooperatives have been initially financed by the Rural Utilities Service (prior Rural Electrification Administration), U.S. Department of Agriculture.

- **coordination service pricing**

The typical price components of a bulk power coordination sale are an energy charge, a capacity, or reservation charge, and an adder. The price for a particular sale may embody some or all of these components. The energy charge is made on a per-kilowatt basis and is intended to recover the seller's system incremental variable costs of making a sale. Because the non-fuel expenses are usually hard to quantify, and small relative to fuel expense, energy charges quoted are usually based on fuel cost. A capacity charge is set at a certain level per kilowatt and is normally paid whether or not energy is taken by the buyer. An adder is added to that energy charge to recover the hard quantify nonfuel variable costs. There are three types of adders: percentage, fixed, and split savings. A percentage adder increases the energy charge by a certain percentage. A fixed adder adds a fixed amount per kilowatt hour to the energy charge. Split savings adders are used only in economy energy transactions. They split production costs savings between the seller and the buyer by adding one half of the savings to the energy cost.

- **coordination service**

The sale, exchange, or transmission of electricity between two or more electric utilities that typically have sufficient generation and transmission capacity to supply their load requirements under normal conditions.

- **cord of wood**

A cord of wood measures 4 feet by 4 feet by 8 feet, or 128 cubic feet.

- **correlation (statistical term)**

In its most general sense, correlation denotes the interdependence between quantitative or qualitative data. It would include the association of dichotomized attributes and the contingency of multiple classified attributes. The concept is quite general and may be extended to more than two variates. The word is most frequently used in a somewhat narrower sense to denote the relationship between measurable variates or ranks.

- **cost model for undiscovered resources**

A computerized algorithm that uses the uranium endowment estimated for a given geological area and selected industry economic indexes to develop random variables that describe the undiscovered resources ultimately expected to be discovered in that area at chosen forward cost categories.

- **cost of capital**

The rate of return a utility must offer to obtain additional funds. The cost of capital varies with the leverage ratio, the effective income tax rate, conditions in the bond and stock markets, growth rate of the utility, its dividend strategy, stability of net income, the amount of new capital required, and other factors dealing with business and financial risks. It is a composite of the cost for debt interest, preferred stock dividends, and common stockholders' earnings that provide the facilities used in supplying utility service.

- **cost of debt**

The interest rate paid on new increments of debt capital multiplied by 1 minus the tax rate.

- **cost of preferred stock**
The preferred stock dividends divided by the net price of the preferred stock.

- **cost of retained earnings**
The residual of an entity's earnings over expenditures, including taxes and dividends, that are reinvested in its business. The cost of these funds is always lower than the cost of new equity capital, due to taxes and transactions costs. Therefore, the cost of retained earnings is the yield that retained earnings accrue upon reinvestment.

- **cost of service**
A ratemaking concept used for the design and development of rate schedules to ensure that the filed rate schedules recover only the cost of providing the electric service at issue. This concept attempts to correlate the utility's costs and revenue with the service provided to each of the various customer classes.

- **cost, insurance, freight (cif)**
A type of sale in which the buyer of the product agrees to pay a unit price that includes the f.o.b. value of the product at the point of origin plus all costs of insurance and transportation. This type of transaction differs from a "delivered" purchase in that the buyer accepts the quantity as determined at the loading port (as certified by the Bill of Loading and Quality Report) rather than pay on the basis of the quantity and quality ascertained at the unloading port. It is similar to the terms of an f.o.b. sale except that the seller, as a service for which he is compensated, arranges for transportation and insurance.

- **cost-of-service regulation**
A traditional electric utility regulation under which a utility is allowed to set rates based on the cost of providing service to customers and the right to earn a limited profit.

- **costs (imports of natural gas)**
All expenses incurred by an importer up to the U.S. point of delivery for the reported quantity imported.

- **criteria pollutant**
A pollutant determined to be hazardous to human health and regulated under EPA's National Ambient Air Quality Standards. The 1970 amendments to the Clean Air Act require EPA to describe the health and welfare impacts of a pollutant as the "criteria" for inclusion in the regulatory regime.

- **crop residue**
Organic residue remaining after the harvesting and processing of a crop.

- **crude oil**
A mixture of hydrocarbons that exists in liquid phase in natural

underground reservoirs and remains liquid at atmospheric pressure after passing through surface separating facilities. Depending upon the characteristics of the crude stream, it may also include: Small amounts of hydrocarbons that exist in gaseous phase in natural underground reservoirs but are liquid at atmospheric pressure after being recovered from oil well (casinghead) gas in lease separators and are subsequently commingled with the crude stream without being separately measured. Lease condensate recovered as a liquid from natural gas wells in lease or field separation facilities and later mixed into the crude stream is also included;

- Small amounts of non-hydrocarbons produced with the oil, such as sulfur and various metals;
- Drip gases, and liquid hydrocarbons produced from tar sands, gilsonite, and oil shale.
- Liquids produced at natural gas processing plants are excluded. Crude oil is refined to produce a wide array of petroleum products, including heating oils; gasoline, diesel and jet fuels; lubricants, asphalt; ethane, propane, and butane; and many other products used for their energy or chemical content.

■ **crude oil acquisitions (unfinished oil acquisitions)**

The volume of crude oil either acquired by the respondent for processing for his own account in accordance with accounting procedures generally accepted and consistently and historically applied by the refiner concerned, or in the case of a processing agreement, delivered to another refinery for processing for the respondent's own account. Crude oil that has not been added by a refiner to inventory and that is thereafter sold or otherwise disposed of without processing for the account of that refiner shall be deducted from its crude oil purchases at the time when the related cost is deducted from refinery inventory in accordance with accounting procedures generally applied by the refiner concerned. Crude oil processed by the respondent for the account of another is not a crude oil acquisition.

■ **crude oil f.o.b. price**

The crude oil price actually charged at the oil producing country's port of loading. Includes deductions for any rebates and discounts or additions of premiums, where applicable. It is the actual price paid with no adjustment for credit terms.

■ **crude oil input**

The total crude oil put into processing units at refineries.

crude oil landed cost

The price of crude oil at the port of discharge, including charges associated with purchasing, transporting, and insuring a cargo from the purchase point to the port of discharge. The cost does not include charges incurred at the discharge port (e.g., import tariffs or fees, wharfage charges, and demurrage).

crude oil less lease condensate

A mixture of hydrocarbons that exists in liquid phase in natural underground reservoirs and remains liquid at atmospheric pressure after passing through surface separating facilities. Such hydrocarbons as lease condensate and natural gasoline recovered as liquids from natural gas wells in lease or field separation facilities and later mixed into the crude stream are excluded. Depending upon the characteristics of the crude stream, crude oil may also include:

1. small amounts of hydrocarbons that exist in gaseous phase in natural underground reservoirs but are liquid at atmospheric pressure after being recovered from oil well (casinghead) gas in lease separators and are subsequently commingled with the crude stream without being separately measured;
2. small amounts of non-hydrocarbons produced with the oil, such as sulfur and various metals.

crude oil losses

Represents the volume of crude oil reported by petroleum refineries as being lost in their operations. These losses are due to spills, contamination, fires, etc., as opposed to refining processing losses.

crude oil production

The volume of crude oil produced from oil reservoirs during given periods of time. The amount of such production for a given period is measured as volumes delivered from lease storage tanks (i.e., the point of custody transfer) to pipelines, trucks, or other media for transport to refineries or terminals with adjustments for (1) net differences between opening and closing lease inventories, and (2) basic sediment and water (BS&W).

crude oil qualities

Refers to two properties of crude oil, the sulfur content, and API gravity, which affect processing complexity and product characteristics.

crude oil refinery input

The total crude oil put into processing units at refineries.

crude oil stocks

Stocks of crude oil and lease condensate held at refineries, in pipelines, at pipeline terminals, and on leases.

- **crude oil used directly**
Crude oil consumed as fuel by crude oil pipelines and on crude oil leases.

- **crude oil, refinery receipts**
Receipts of domestic and foreign crude oil at a refinery. Includes all crude oil in transit except crude oil in transit by pipeline. Foreign crude oil is reported as a receipt only after entry through customs. Crude oil of foreign origin held in bonded storage is excluded.

- **crystalline fully refined wax**
A light colored paraffin wax having the following characteristics: viscosity at 210 degrees Fahrenheit (D88)-59.9 SUS (10.18 centistokes) maximum; oil content (D721)-0.5 percent maximum; other +20 color, Saybolt minimum.

- **crystalline other wax**
A paraffin wax having the following characteristics: viscosity at 210 deg. F(D88)-59.9 SUS (10.18 centistokes) maximum; oil content (D721)-0.51 percent minimum to 15 percent maximum.

- **cubic foot (cf), natural gas**
The amount of natural gas contained at standard temperature and pressure (60 degrees Fahrenheit and 14.73 pounds standard per square inch) in a cube whose edges are one foot long.

- **cull wood**
Wood logs, chips, or wood products that are burned.

- **culm**
Waste from Pennsylvania anthracite preparation plants, consisting of coarse rock fragments containing as much as 30 percent small sized coal; sometimes defined as including very fine coal particles called silt. Its heat value ranges from 8 to 17 million Btu per short ton.

- **cultivar**
A horticulturally or agriculturally derived variety of a plant.

- **cumulative depletion**
The sum in tons of coal extracted and lost in mining as of a stated date for a specified area or a specified coal bed.

- **current (electric)**
A flow of electrons in an electrical conductor. The strength or rate of movement of the electricity is measured in amperes.

- **current assets**
Cash and other assets that are expected to be turned into cash, sold, or exchanged within the normal operating cycle of the utility, usually one year. Current assets include cash, marketable securities, receivables, inventory, and current prepayments.

- **current liabilities**

A debt or other obligation that must be discharged within one year or the normal operating cycle of the utility by expending a current asset or the incurrence of another short-term obligation. Current liabilities include accounts payable, short-term notes payable, and accrued expenses payable such as taxes payable and salaries payable.

- **current ratio**

The ratio of current assets divided by current liabilities that shows the ability of a utility to pay its current obligations from its current assets. A measure of liquidity, the higher the ratio, the more assurance that current liabilities can be paid.

- **customer choice**

The right of customers to purchase energy from a supplier other than their traditional supplier or from more than one seller in the retail market.

- **customs district (coal)**

Customs districts, as defined by the Bureau of the Census, U.S. Department of Commerce, "Monthly Report EM 545," are as follows:

- **Eastern**: Bridgeport, CT, Washington, DC, Boston, MA, Baltimore, MD, Portland, ME, Buffalo, NY, New York City, NY, Ogdensburg, NY, Philadelphia, PA, Providence, RI, Norfolk, VA, St. Albans, VT.
- **Southern**: Mobile, AL, Savannah, GA, Miami, FL, Tampa, FL, New Orleans, LA, Wilmington, NC, San Juan, PR, Charleston, SC, Dallas-Fort Worth, TX, El Paso, TX, Houston-Galveston, TX, Laredo, TX, Virgin Islands.
- **Western**: Anchorage, AK, Nogales, AZ, Los Angeles, CA, San Diego, CA, San Francisco, CA, Honolulu, HI, Great Falls, MT, Portland, OR, Seattle, WA.
- **Northern**: Chicago, IL, Detroit, MI, Duluth, MN, Minneapolis, MN, St. Louis, MO, Pembina, ND, Cleveland, OH, Milwaukee, WI.

- **cut-off grade (uranium)**

The lowest grade, in percent U_3O_8, of uranium ore at a minimum specified thickness that can be mined at a specified cost.

- **cycle**

The time period running from the startup of one reactor cycle to the startup of the following cycle.

- **cycle/reactor history**

A group of assemblies that have been irradiated in the same cycles in an individual reactor and are said to have the same cycle/reactor history.

- **cycling (natural gas)**

The practice of producing natural gas for the extraction of natural gas liquids, returning the dry residue to the producing reservoir to maintain reservoir pressure and increase the ultimate recovery of natural gas liquids. The reinjected gas is produced for disposition after cycling operations are completed.

- **dam**

A physical barrier constructed across a river or waterway to control the flow of or raise the level of water. The purpose of construction may be for flood control, irrigation needs, hydroelectric power production, and/or recreation usage.

- **day lighting controls**

A system of sensors that assesses the amount of daylight and controls lighting or shading devices to maintain a specified lighting level. The sensors are sometimes referred to as "photocells."

- **day-ahead and hour-ahead markets**

Forward markets where electricity quantities and market clearing prices are calculated individually for each hour of the day on the basis of participant bids for energy sales and purchases.

- **day-ahead schedule**

A schedule prepared by a scheduling coordinator or the independent system operator before the beginning of a trading day. This schedule indicates the levels of generation and demand scheduled for each settlement period that trading day.

- **deadweight tons**

The lifting capacity of a ship expressed in long tons (2,240 lbs.), including cargo, commodities, and crew.

- **dealer tank wagon (dtw) sales**

Wholesale sales of gasoline priced on a delivered basis to a retail outlet.

- **decatherm**

Ten therms or 1,000,000 Btu.

- **decommissioning**

Retirement of a nuclear facility, including decontamination and/or dismantlement.

- **decontamination**

Removal of unwanted radioactive or hazardous contamination by a chemical or mechanical process.

- **dedicated reserves**

The volume of recoverable, salable gas reserves committed to, controlled by, or possessed by the reporting pipeline company and used for acts and services for which both the seller and the company have received certificate authorization from the Federal Energy Regulatory Commission (FERC). Reserves include both company-owned reserves (including owned gas in underground storage), reserves under contract from independent producers, and short-term and emergency supplies from the intrastate market. Gas volumes under contract from other interstate pipelines are not included as reserves, but may constitute part or all of a company's gas supply.

- **dedicated vehicle**
  A vehicle that operates only on an alternative fuel, as when a vehicle is configured to operate on compressed natural gas.

- **deepest total depth**
  The deepest total depth of a given well is the distance from a surface reference point (usually the Kelly bushing) to the point of deepest penetration measured along the well bore. If a well is drilled from a platform or barge over water, the depth of the water is included in the total length of the well bore.

- **deferred cost**
  An expenditure not recognized as a cost of operation of the period in which incurred, but carried forward to be written off in future periods.

- **deferred fuel costs**
  An expenditure for fuel that is not recognized for bookkeeping practices as a cost in the operating period incurred, but carried forward to be written off in future periods.

- **deferred income tax (liability)**
  A liability in the balance sheet representing the additional Federal income taxes that would have been due if a utility had not been allowed to compute tax expenses differently for income tax reporting purposes than for ratemaking purposes.

- **deforestation**
  The net removal of trees from forested land.

- **degasification system**
  The methods employed for removing methane from a coal seam that could not otherwise be removed by standard ventilation fans and thus would pose a substantial hazard to coal miners. These systems may be used prior to mining or during mining activities.

- **degradable organic carbon**
  The portion of organic carbon present in such solid waste as paper, food waste, and yard waste that is susceptible to biochemical decomposition.

- **delayed coking**
  A process by which heavier crude oil fractions can be thermally decomposed under conditions of elevated temperatures and pressure to produce a mixture of lighter oils and petroleum coke. The light oils can be processed further in other refinery units to meet product specifications. The coke can be used either as a fuel or in other applications such as the manufacturing of steel or aluminum.

■ **delayed coking**
A process by which heavier crude oil fractions can be thermally decomposed under conditions of elevated temperatures and pressure to produce a mixture of lighter oils and petroleum coke. The light oils can be processed further in other refinery units to meet product specifications. The coke can be used either as a fuel or in other applications such as the manufacturing of steel or aluminum.

■ **deliverability**
Represents the number of future years during which a pipeline company can meet its annual requirements for its presently certificated delivery capacity from presently committed sources of supply. The availability of gas from these sources of supply shall be governed by the physical capabilities of these sources to deliver gas by the terms of existing gas-purchase contracts, and by limitations imposed by State or Federal regulatory agencies.

■ **delivered (gas)**
The physical transfer of natural, synthetic, and/or supplemental gas from facilities operated by the responding company to facilities operated by others or to consumers.

■ **delivered cost**
The cost of fuel, including the invoice price of fuel, transportation charges, taxes, commissions, insurance, and expenses associated with leased or owned equipment used to transport the fuel.

■ **delivered energy**
The amount of energy delivered to the site (building); no adjustment is made for the fuels consumed to produce electricity or district sources. This is also referred to as net energy.

■ **deliveries (electric)**
Energy generated by one system and delivered to another system through one or more transmission lines.

■ **demand bid**
A bid into the power exchange indicating a quantity of energy or an ancillary service that an eligible customer is willing to purchase and, if relevant, the maximum price that the customer is willing to pay.

■ **demand charge credit**
Compensation received by the buyer when the delivery terms of the contract cannot be met by the seller.

■ **demand charge**
That portion of the consumer's bill for electric service based on the consumer's maximum electric capacity usage and calculated based on the billing demand charges under the applicable rate schedule.

- **demand indicator**

  A measure of the number of energy-consuming units, or the amount of service or output, for which energy inputs are required.

- **demand interval**

  The time period during which flow of electricity is measured (usually in 15-, 30-, or 60-minute increments.)

- **demand-metered**

  Having a meter to measure peak demand (in addition to total consumption) during a billing period. Demand is not usually metered for other energy sources.

- **demand-side management (dsm)**

  The planning, implementation, and monitoring of utility activities designed to encourage consumers to modify patterns of electricity usage, including the timing and level of electricity demand. It refers to only energy and load-shape modifying activities that are undertaken in response to utility-administered programs. It does not refer to energy and load-shaped changes arising from the normal operation of the marketplace or from government -mandated energy-efficiency standards. Demand- Side Management covers the complete range of load-shape objectives, including strategic conservation and load management, as well as strategic load growth.

- **demand-side management costs**

  The costs incurred by the utility to achieve the capacity and energy savings from the Demand-Side Management Program. Costs incurred by customers or third parties are to be excluded. The costs are to be reported in thousands of dollars (nominal) in the year in which they are incurred, regardless of when the savings occur. The utility costs are all the annual expenses (labor, administrative, equipment, incentives, marketing, monitoring and evaluation, and other incurred by the utility for operation of the DSM Program), regardless of whether the costs are expensed or capitalized. Lump sum capital costs (typically accrued over several years prior to start up) are not to be reported. Program costs associated with strategic load growth activities are also to be excluded.

- **demonstrated reserve base (coal)**

  A collective term for the sum of coal in both measured and indicated resource categories of reliability, representing 100 percent of the in-place coal in those categories as of a certain date. Includes beds of bituminuous coal and anthracite 28 or more inches thick and beds of subbituminuous coal 60 or more inches thick that can occur at depths of up to 1,000 feet. Includes beds of lignite 60 or more inches thick that can be surface mined. Includes also thinner and/or deeper beds that presently are being mined or for which there is evidence that they could be mined commercially at a given time. Represents that portion of the identified coal resource from which reserves are calculated.

- **demonstrated resources**

Same qualifications as identified resources, but include measured and indicated degrees of geologic assurance and excludes the inferred.

- **demonstration and test vehicles**

Vehicles operated by a motor vehicle dealer solely for the purpose of promoting motor vehicle sales or permitting potential purchasers to drive the vehicle for pre-purchase or pre-lease evaluation; or a vehicle that is owned and operated by a motor vehicle manufacturer or motor vehicle component manufacturer, or owned or held by a university research department, independent testing laboratory, or other such evaluation facility, solely for the purpose of evaluating the performance of such vehicles for engineering, research and development, or quality control reasons.

- **demurrage**

The charge paid to the vessel owner or operator for detention of a vessel at the port(s) beyond the time allowed, usually 72 hours, for loading and unloading.

- **dependable capacity**

The load-carrying ability of a station or system under adverse conditions for a specified period of time.

- **depleted resources**

Resources that have been mined; include coal recovered, coal lost in mining, and coal reclassified as sub-economic because of mining.

- **depleted storage field**

A sub-surface natural geological reservoir, usually a depleted gas or oil field, used for storing natural gas.

- **depletion (coal)**

The subtraction of both tonnage produced and the tonnage lost to mining from identified resources to determine the remaining tonnage as of a certain time.

- **depletion allowance**

A term for either (1) a periodic assignment to expense of recorded amounts or (2) an allowable income tax deduction that is related to the exhaustion of mineral reserves. Depletion is included as one of the elements of amortization. When used in that manner, depletion refers only to book depletion.

- **Book.** The portion of the carrying value (other than the portion associated with tangible assets) prorated in each accounting period, for financial reporting purposes, to the extracted portion of an economic interest in wasting natural resource.
- **Tax-cost.** A deduction (allowance) under U.S. Federal income taxation normally calculated under a formula whereby the adjusted basis of the mineral property is multiplied by a fraction, the numerator of which is the number of units of minerals sold during the tax year and the denominator of which is the estimated number of units of unextracted minerals remaining at the end of the tax year plus the number of units of minerals sold during the tax year.
- **Tax-percentage (for Statutory).** A deduction (allowance) allowed to certain mineral producers under U.S. Federal income taxation calculated on the basis of a specified percentage of gross revenue from the sale of minerals from each mineral property not to exceed the lesser of 50 percent of the taxable income from the property computed without allowance for depletion. (There are also other limits on percentage depletion on oil and gas production.) The taxpayer is entitled to a deduction representing the amount of tax-cost depletion or percentage (statutory) depletion, whichever is higher.
- **Excess statutory depletion.** The excess of estimated statutory depletion allowable as an income tax deduction over the amount of cost depletion otherwise allowable as a tax deduction, determined on a total enterprise basis.

■ **depletion factor**

The multiplier applied to the tonnage produced to compute depletion. This multiplier takes into account both the tonnage recovered and the tonnage lost due to mining. The depletion factor is the reciprocal of the recovery factor in relation to a given quantity of production.

■ **depreciation and amortization of property, plant, and equipment**

The monthly provision for depreciation and amortization (applicable to utility property other than electric plant, electric plant in service, and equipment).

■ **depth of deepest production**

The depth of the deepest production is the length of the well bore measured (in feet) from the surface reference point to the bottom of the open hole or the deepest perforation in the casing of a producing well.

- **deregulation**
The elimination of some or all regulations from a previously regulated industry or sector of an industry.

- **design electrical rating (capacity) net**
The nominal net electrical output of a nuclear unit, as specified by the utility for the purpose of plant design.

- **design head**
The achieved river, pondage, or reservoir surface height (forebay elevation) that provides the water level to produce the full flow at the gate of the turbine in order to attain the manufacturer's installed nameplate rating for generation capacity.

- **desulfurization**
The removal of sulfur, as from molten metals, petroleum oil, or flue gases.

- **development costs**
Costs incurred to obtain access to proved reserves and to provide facilities for extracting, treating, gathering, and storing the oil and gas. More specifically, development costs, depreciation and applicable operating costs of support equipment and facilities, and other costs of development activities, are costs incurred to:

- **development drilling**
Drilling done to determine more precisely the size, grade, and configuration of an ore deposit subsequent to when the determination is made that the deposit can be commercially developed. Not included are: (1) secondary development drilling, (2) solution-mining drilling for production, or (3) production-related underground and open pit drilling done for control of mining operations.

- **development**
The preparation of a specific mineral deposit for commercial production; this preparation includes construction of access to the deposit and of facilities to extract the minerals. The development process is sometimes further distinguished between a preproduction stage and a current stage, with the distinction being made on the basis of whether the development work is performed before or after production from the mineral deposit has commenced on a commercial scale.

- **development well**
A well drilled within the proved area of an oil or gas reservoir to the depth of a stratigraphic horizon known to be productive.

▪ **diesel fuel**

A fuel composed of distillates obtained in petroleum refining operation or blends of such distillates with residual oil used in motor vehicles. the boiling point and specific gravity are higher for diesel fuels than for gasoline.

▪ **diesel fuel system**

Diesel engines are internal combustion engines that burn diesel oil rather than gasoline. Injectors are used to spray droplets of diesel oil into the combustion chambers, at or near the top of the compression stroke. Ignition follows due to the very high temperature of the compressed intake air, or to the use of "glow plugs," which retain heat from previous ignitions (spark plugs are not used). Diesel engines are generally more fuel-efficient than gasoline engines but must be stronger and heavier because of high compression ratios.

▪ **diesel-electric plant**

A generating station that uses diesel engines to drive its electric generators.

▪ **diffusive transport**

The process by which particles of liquids or gases move from an area of higher concentration to an area of lower concentration.

▪ **direct access**

The ability of a retail customer to purchase electricity or other energy sources directly from a supplier other than their traditional supplier.

▪ **direct control load management**

The magnitude of customer demand that can be interrupted at the time of the seasonal peak load by direct control of the system operator by interrupting power supply to individual appliances or equipment on customer premises. This type of control usually reduces the demand of residential customers.

▪ **direct electricity load control**

The utility installs a radio-controlled device on the HVAC equipment. During periods of particularly heavy use of electricity, the utility will send a radio signal to the building in its service territory with this device and turn off the HVAC for a certain period.

▪ **direct labor hours**

Direct labor hours worked by all mining employees at a mining operation during the year. Includes hours worked by those employees engaged in production, preparation, development, maintenance, repair, shop or yard work management, and technical or engineering work. Excludes office workers. Excludes vacation and leave hours.

- **direct load control**

This Demand-Side Management category represents the consumer load that can be interrupted at the time of annual peak load by direct control of the utility system operator. Direct Load Control does not include Interruptible Load. This type of control usually involves residential consumers.

- **direct milling cost**

Operating costs directly attributable to the processing of ores or other feed materials, including labor, supervision, engineering, power, fuel, supplies, reagents, and maintenance.

- **direct mining cost**

Operating cost directly attributable to the mining of ore, including costs for labor, supervision, engineering, power, fuel, supplies, equipment replacement, maintenance, and taxes on production.

- **direct nonprocess end use**

Those end uses that may be found on commercial, residential, or other sites, as well as at manufacturing establishments. They include heating, ventilation, and air conditioning (HVAC), facility lighting, facility support, onsite transportation, conventional electricity generation, and other nonprocess uses. "Direct" denotes that only the quantities of electricity or fossil fuel used in their original state (i.e., not transformed) are included in the estimates.

- **direct process end use**

Those end uses that are specific to the carrying out of manufacturing. They include process heating, process cooling and refrigeration, machine drive, electrochemical processes, and other process uses. "Direct" denotes that only the quantities of electricity or fossil fuel used in their original state (i.e., not transformed) are included in the estimates.

- **direct use**

Use of electricity that 1) is self-generated, 2) is produced by either the same entity that consumes the power or an affiliate, and 3) is used in direct support of a service or industrial process located within the same facility or group of facilities that house the generating equipment. Direct use is exclusive of station use.

- **direct utility cost**

A utility cost that is identified with one of the DSM program categories (e.g. Energy Efficiency or Load Management).

- **directional (deviated) well**

A well purposely deviated from the vertical, using controlled angles to reach an objective location other than directly below the surface location. A directional well may be the original hole or a directional "sidetrack" hole that deviates from the original bore at some point below the surface. The new footage associated with directional "sidetrack" holes should not be confused with footage resulting from remedial sidetrack drilling. If there is a common bore from which two or more wells are drilled, the first complete bore from the surface to the original objective is classified and reported as a well drilled. Each of the deviations from the common bore is reported as a separate well.

- **discharged fuel**

Irradiated fuel removed from a nuclear reactor during refueling.

- **discrete-delivery energy sources**

Energy sources that must be delivered to a site.

- **dispatching**

The operating control of an integrated electric system involving operations such as (1) the assignment of load to specific generating stations and other sources of supply to effect the most economical supply as the total or the significant area loads rise or fall (2) the control of operations and maintenance of high-voltage lines, substations, and equipment; (3) the operation of principal tie lines and switching; (4) the scheduling of energy transactions with connecting electric utilities.

- **disposition, natural gas**

The removal of natural, synthetic, and/or supplemental gas, or any components or gaseous mixtures contained therein, from the responding company's facilities within the report State by any means or for any purpose, including the transportation of such gas out of the report State.

- **disposition, petroleum**

A set of categories used to account for how crude oil and petroleum products are transferred, distributed, or removed from the supply stream. The categories include stock change, crude oil losses, refinery inputs, exports, and products supplied for domestic consumption.

- **distillate fuel oil**

A general classification for one of the petroleum fractions produced in conventional distillation operations. It includes diesel fuels and fuel oils. Products known as No. 1, No. 2, and No. 4 diesel fuel are used in on-highway diesel engines, such as those in trucks and automobiles, as well as off-highway engines, such as those in railroad locomotives and agricultural machinery. Products known as No. 1, No. 2, and No. 4 fuel oils are used primarily for space heating and electric power generation.

No. 1 Distillate: A light petroleum distillate that can be used as either a diesel fuel or a fuel oil.

No. 1 Diesel Fuel: A light distillate fuel oil that has distillation temperatures of 550 degrees Fahrenheit at the 90-percent point and meets the specifications defined in ASTM Specification D 975. It is used in high-speed diesel engines, such as those in city buses and similar vehicles.

No. 1 Fuel Oil: A light distillate fuel oil that has distillation temperatures of 400 degrees Fahrenheit at the 10-percent recovery point and 550 degrees Fahrenheit at the 90-percent point and meets the specifications defined in ASTM Specification D 396. It is used primarily as fuel for portable outdoor stoves and portable outdoor heaters.

No. 2 Distillate: A petroleum distillate that can be used as either a diesel fuel or a fuel oil.

No. 2 Diesel Fuel: A fuel that has distillation temperatures of 500 degrees Fahrenheit at the 10-percent recovery point and 640 degrees Fahrenheit at the 90-percent recovery point and meets the specifications defined in ASTM Specification D 975. It is used in high-speed diesel engines, such as those in railroad locomotives, trucks, and automobiles.

Low Sulfur No. 2 Diesel Fuel: No. 2 diesel fuel that has a sulfur level no higher than 0.05 percent by weight. It is used primarily in motor vehicle diesel engines for on-highway use.

High Sulfur No. 2 Diesel Fuel: No. 2 diesel fuel that has a sulfur level above 0.05 percent by weight.

No. 2 Fuel oil (Heating Oil): A distillate fuel oil that has distillation temperatures of 400 degrees Fahrenheit at the 10-percent recovery point and 640 degrees Fahrenheit at the 90-percent recovery point and meets the specifications defined in ASTM Specification D 396. It is used in atomizing type burners for domestic heating or for moderate capacity commercial/industrial burner units.

No. 4 Fuel: A distillate fuel oil made by blending distillate fuel oil and residual fuel oil stocks. It conforms with ASTM Specification D 396 or Federal Specification VV-F-815C and is used extensively in industrial plants and in commercial burner installations that are not equipped with preheating facilities. It also includes No. 4 diesel fuel used for low- and medium-speed diesel engines and conforms to ASTM Specification D 975.

▪ distillation unit (atmospheric)

The primary distillation unit that processes crude oil (including mixtures of other hydrocarbons) at approximately atmospheric conditions. It includes a pipe still for vaporizing the crude oil and a fractionation tower for separating the vaporized hydrocarbon components in the crude oil into fractions with different boiling ranges. This is done by continuously vaporizing and condensing the components to separate higher oiling point material. The selected boiling ranges are set by the processing scheme, the properties of the crude oil, and the product specifications.

- **distributed/point-of-use water-heating system**

A system for heating hot water, for other than space heating purposes, which is located at more than one space within a building. A point-of-use water heater is located at the faucet and heats water only as required for immediate use. Because water is not heated until it is required, this equipment is more energy-efficient.

- **distribution system**

The portion of the transmission and facilities of an electric system that is dedicated to delivering electric energy to an end-user.

- **distribution**

The delivery of energy to retail customers.

- **distribution use**

Natural gas used as fuel in the respondent's operations.

- **distributor**

A company primarily engaged in the sale and delivery of natural and/or supplemental gas directly to consumers through a system of mains.

- **district chilled water**

Chilled water from an outside source used as an energy source for cooling in a building. The water is chilled in a central plant and piped into the building. Chilled water may be purchased from a utility or provided by a central physical plant in a separate building that is part of the same multi-building facility (for example, a hospital complex or university).

- **district heat**

Steam or hot water from an outside source used as an energy source in a building The steam or hot water is produced in a central plant and piped into the building. The district heat may be purchased from a utility or provided by a physical plant in a separate building that is part of the same facility (for example, a hospital complex or university).

- **diversity exchange**

An exchange of capacity or energy, or both, between systems whose peak loads occur at different times.

- **diversity**

The electric utility system's load is made up of many individual loads that make demands upon the system usually at different times of the day. The individual loads within the customer classes follow similar usage patterns, but these classes of service place different demands upon the facilities and the system grid. The service requirements of one electrical system can differ from another by time-of-day usage, facility usage, and/or demands placed upon the system grid.

- **divestiture**
The stripping off of one utility function from the others by selling (spinning-off) or in some other way changing the ownership of the assets related to that function. Stripping off is most commonly associated with spinning-off generation assets so they are no longer owned by the shareholders that own the transmission and distribution assets.

- **docket**
A formal record of a Federal Energy Regulatory Commission proceeding. These records are available for inspection and copying by the public. Each individual case proceeding is identified by an assigned number.

- **doe**
Department of Energy.

- **domestic inland consumption**
Domestic inland consumption is the sum of all refined petroleum products supplied for domestic use (excludes international marine bunkers). Consumption is calculated by product by adding production, imports, crude oil burned directly, and refinery fuel and losses, and then subtracting exports and charges in primary stocks (net withdrawals is a plus quantity and net additions is a minus quantity).

- **domestic uranium industry**
Collectively, those businesses (whether U.S. or foreign-based) that operate under the laws and regulations pertaining to the conduct of commerce within the United States and its territories and possessions and that engage in activities within the United States, its territories, and possessions specifically directed toward uranium exploration, development, mining, and milling; marketing of uranium materials; enrichment; fabrication; or acquisition and management of uranium materials for use in commercial nuclear power plants.

- **domestic vehicle producer**
An Original Vehicle Manufacturer that assembles vehicles in the United States for domestic use. The term "domestic" pertains to the fifty states, the District of Columbia, commonwealths, territories, and possessions of the United States.

- **double circuit line**

A transmission line having two separate circuits. Because each carries three-phase power, at least six conductors, three per circuit, are required.

- **drainage basin**

The land drained by a river system.

- **drawdown (maximum)**

The distance that the water surface of the reservoir is lowered from the normal full elevation to the lowest allowable elevation as the result of the withdrawal of water for the purposes of generating electricity.

- **drawdown**

The lowering of the water level of a reservoir as a result of withdrawing water.

- **dredge mining**

A method of recovering coal from rivers or streams.

- **drift mine**

A mine that opens horizontally into the coal bed or coal outcrop. Drill and equip development wells, development-type stratigraphic test wells, and service wells, including the costs of platforms and of well equipment such as casing, tubing, pumping equipment, and the wellhead assembly.

- **drilling and equipping of wells**

The drilling and equipping of wells through completion of the "Christmas tree."

- **drilling arrangement**

A contractual agreement under which a working interest owner (assignor) assigns a part of a working interest in a property to another party (the assignee) in exchange for which the assignee agrees to develop the property. The term may also be applied to an agreement under which an operator assigns fractional shares in production from a property to participants for cash considerations as a means of acquiring cash for developing the property.

Under a "disproportionate cost" drilling arrangement, the participants normally pay a greater total share of costs than the total value of the fractional shares of the property received in the arrangement.

- **drilling**

The act of boring a hole (1) to determine whether minerals are present in commercially recoverable quantities and (2) to accomplish production of the minerals (including drilling to inject fluids).

  - **Exploratory.** Drilling to locate probable mineral deposits or to establish the nature of geological structures; such wells may not be capable of production if minerals are discovered.
  - **Developmental.** Drilling to delineate the boundaries of a

known mineral deposit to enhance the productive capacity of the producing mineral property.

- **Directional.** Drilling that is deliberately made to depart significantly from the vertical.

dry (coal) basis

Coal quality data calculated to a theoretical basis in which no moisture is associated with the sample. This basis is determined by measuring the weight loss of a sample when its inherent moisture is driven off under controlled conditions of low temperature air-drying followed by heating to just above the boiling point of water (104 to 110 degrees Centigrade).

dry bottom boiler

No slag tanks at furnace throat area. The throat area is clear. Bottom ash drops through the throat to the bottom ash water hoppers. This design is used where the ash melting temperature is greater than the temperature on the furnace wall, allowing for relatively dry furnace wall conditions.

dry hole

An exploratory or development well found to be incapable of producing either oil or gas in sufficient quantities to justify completion as an oil or gas well.

dry hole charge

The charge-off to expense of a previously capitalized cost upon the conclusion of an unsuccessful drilling effort.

dry hole contribution

A payment (either in cash or acreage) that is required by agreement only if a test well is unsuccessful and that is made in exchange for well test and evaluation data.

dry natural gas

Natural gas which remains after: 1) the liquefiable hydrocarbon portion has been removed from the gas stream (i.e., gas after lease, field, and/or plant separation); and 2) any volumes of nonhydrocarbon gases have been removed where they occur in sufficient quantity to render the gas unmarketable. Note: Dry natural gas is also known as consumer-grade natural gas. The parameters for measurement are cubic feet at 60 degrees Fahrenheit and 14.73 pounds per square inch absolute.

dry natural gas production

The process of producing consumer-grade natural gas. Natural gas withdrawn from reservoirs is reduced by volumes used at the production (lease) site and by processing losses. Volumes used at the production site include (1) the volume returned to reservoirs in cycling, repressuring of oil reservoirs, and conservation operations; and (2) gas vented and flared. Processing losses include

(1) nonhydrocarbon gases (e.g., water vapor, carbon dioxide, helium, hydrogen sulfide, and nitrogen) removed from the gas stream; and (2) gas converted to liquid form, such as lease condensate and plant liquids. Volumes of dry gas withdrawn from gas storage reservoirs are not considered part of production. Dy natural gas production equals marketed production less extraction loss.

- **dual fuel vehicle (1)**

A motor vehicle that is capable of operating on an alternative fuel and on gasoline or diesel fuel. These vehicles have at least two separate fuel systems which inject each fuel simultaneously into the engine combustion chamber.

- **dual fuel vehicle (2)**

A motor vehicle that is capable of operating on an alternative fuel and on gasoline or diesel fuel. This term is meant to represent all such vehicles whether they operate on the alternative fuel and gasoline/ diesel simultaneously (e.g., flexible-fuel vehicles) or can be switched to operate on gasoline/ diesel or an alternative fuel (e.g., bi-fuel vehicles).

- **dual-fired unit**

A generating unit that can produce electricity using two or more input fuels. In some of these units, only the primary fuel can be used continuously; the alternate fuel(s) can be used only as a start-up fuel or in emergencies.

- **dump energy**

Energy generated in a hydroelectric plant by water that cannot be stored or conserved and which energy is in excess of the needs of the system producing the energy.

- **e85**

A fuel containing a mixture of 85 percent ethanol and 15 percent gasoline.

- **e95**

A fuel containing a mixture of 95 percent ethanol and 5 percent gasoline

- **economy of scale**

The principle that larger production facilities have lower unit costs than smaller facilities.

- **effective full-power days**

The number of effective full-power days produced by a unit is a measure of the unit's energy generation. It is determined using the following ratio: Heat generation (planned or actual) in megawatt days thermal (MWdt)(divided by) Licensed thermal power in megawatts thermal (MWt)

- **eia**

The Energy Information Administration. An independent agency within the U.S. Department of Energy that develops surveys, collects energy data, and analyzes and models energy issues. The Agency must meet the requests of Congress, other elements within the Department of Energy, Federal Energy Regulatory Commission, the Executive Branch, its own independent needs, and assist the general public, or other interest groups, without taking a policy position.

- **electric baseboard**
An individual space heater with electric resistance coils mounted behind shallow panels along baseboards. Electric baseboards rely on passive convection to distribute heated air to the space.

- **electric current**
The flow of electric charge. The preferred unit of measure is the ampere.

- **electric energy**
The ability of an electric current to produce work, heat, light, or other forms of energy. It is measured in kilowatt hours.

- **electric expenses**
The cost of labor, material, and expenses incurred in operating a facility's prime movers, generators, auxiliary apparatus, switching gear, and other electric equipment for each of the points where electricity enters the transmission or distribution grid.

- **electric generation industry**
Stationary and mobile generating units that are connected to the electric power grid and can generate electricity. The electric generation industry includes the "electric power sector" (utility generators and independent power producers) and industrial and commercial power generators, including combined-heat-and-power producers, but excludes units at single-family dwellings.

- **electric generator**
A facility that produces only electricity, commonly expressed in kilowatt hours (kWh) or megawatt hours (MWh). Electric generators include electric utilities and independent power producers.

- **electric hybrid vehicle**
An electric vehicle that either (1) operates solely on electricity, but contains an internal combustion motor that generates additional electricity (series hybrid); or (2) contains an electric system and an internal combustion system and is capable of operating on either system (parallel hybrid).

- **electric industry reregulation**
The design and implementation of regulatory practices to be applied to the remaining traditional utilities after the electric power industry has been restructured.

Reregulation applies to those entities that continue to exhibit characteristics of a natural monopoly. Reregulation could employ the same or different regulatory practices as those used before restructuring.

- **electric industry restructuring**

The process of replacing a monopolistic system of electric utility suppliers with competing sellers, allowing individual retail customers to choose their supplier but still receive delivery over the power lines of the local utility. It includes the reconfiguration of vertically-integrated electric utilities.

- **electric motor vehicle**

A motor vehicle powered by an electric motor that draws current from rechargeable storage batteries, fuel cells, photovoltaic arrays, or other sources of electric current.

- **electric operating expenses**

Summation of electric operation-related expenses, such as operation expenses, maintenance expenses, depreciation expenses, amortization, taxes other than income taxes, Federal income taxes, other income taxes, provision for deferred income taxes, provision for deferred income-credit, and investment tax credit adjustment.

- **electric plant (physical)**

A facility containing prime movers, electric generators, and auxiliary equipment for converting mechanical, chemical, and/or fission energy into electric energy.

- **electric plant acquisition adjustment**

The difference between (a) the cost to the respondent utility of an electric plant acquired as an operating unit or system by purchase and (b) the depreciated original cost, estimated if not known, of such property.

- **electric power grid**

A system of synchronized power providers and consumers connected by transmission and distribution lines and operated by one or more control centers. In the continental United States, the electric power grid consists of three systems: the Eastern Interconnect, the Western Interconnect, and the Texas Interconnect. In Alaska and Hawaii, several systems encompass areas smaller than the State (e.g., the interconnect serving Anchorage, Fairbanks, and the Kenai Peninsula; individual islands).

- **electric power plant**

A station containing prime movers, electric generators, and auxiliary equipment for converting mechanical, chemical, and/or fission energy into electric energy.

- **electric power sector**

An energy-consuming sector that consists of electricity only and combined heat and power(CHP) plants whose primary business is to sell electricity, or electricity and heat, to the public—i.e., North American Industry Classification System 22 plants.

- **electric power system**

An individual electric power entity—a company; an electric cooperative; a public electric supply corporation as the Tennesse Valley Authority; a similar Federal department or agency such as the Bonneville Power Administration; the Bureau of Reclamation or the Corps of Engineers; a municipally owned electric department offering service to the public; or an electric public utility district (a "PUD"); also a jointly owned electric supply project such as the Keystone.

- **electric power**

The rate at which electric energy is transferred. Electric power is measured by capacity and is commonly expressed in megawatts (MW).

- **electric pump for well water**

This pump forces the water from a well below ground level up into the water pipes that circulate through the house. When this pump is not working, there is a limited supply of running water in the house.

- **electric rate schedule**

A statement of the electric rate and the terms and conditions governing its application, including attendant contract terms and conditions that have been accepted by a regulatory body with appropriate oversight authority.

- **electric rate**

The price set for a specified amount and type of electricity by class of service in an electric rate schedule or sales contract.

- **electric system loss**

Total energy loss from all causes for an electric utility.

- **electric system reliability**

The degree to which the performance of the elements of the electrical system results in power being delivered to consumers within accepted standards and in the amount desired. Reliability encompasses two concepts, adequacy and security. Adequacy implies that there are sufficient generation and transmission resources installed and available to meet projected electrical demand plus reserves for contingencies. Security implies that the system will remain intact operationally (i.e., will have sufficient available operating capacity) even after outages or other equipment failure. The degree of reliability may be measured by the frequency, duration, and magnitude of adverse effects on consumer service.

- **electric utility**
  A corporation, person, agency, authority, or other legal entity or instrumentality aligned with distribution facilities for delivery of electric energy for use primarily by the public. Included are investor-owned electric utilities, municipal and State utilities, Federal electric utilities, and rural electric cooperatives. A few entities that are tariff based and corporately aligned with companies that own distribution facilities are also included.

- **electric utility divestiture**
  The separation of one electric utility function from others through the selling of the management and ownership of the assets related to that function. It is most commonly associated with selling generation assets so they are no longer owned or controlled by the shareholders that own the company's transmission and distribution assets.

- **electric utility restructuring**
  The introduction of competition into at least the generation phase of electricity production, with a corresponding decrease in regulatory control.

- **electric utility sector**
  The electric utility sector consists of privately and publicly owned establishments that generate, transmit, distribute, or sell electricity primarily for use by the public and that meet the definition of an electric utility. Nonutility power producers are not included in the electric sector.

- **electric zone**
  A portion of the grid controlled by the independent system operator.

- **electrical system energy losses**
  The amount of energy lost during generation, transmission, and distribution of electricity, including plant and unaccounted for use.

- **electricity**
  A form of energy characterized by the presence and motion of elementary charged particles generated by friction, induction, or chemical change.

- **electricity broker**
  An entity that arranges the sale and purchase of electric energy, the transmission of electricity, and/or other related services between buyers and sellers but does not take title to any of the power sold.

- **electricity congestion**
  A condition that occurs when insufficient transmission capacity

is available to implement all of the desired transactions simultaneously.

- **electricity demand bid**
A bid into the power exchange indicating a quantity of energy or an ancillary service that an eligible customer is willing to purchase and, if relevant, the maximum price that the customer is willing to pay.

- **electricity demand**
The rate at which energy is delivered to loads and scheduling points by generation, transmission, and distribution facilities.

- **electricity generation**
The process of producing electric energy or the amount of electric energy produced by transforming other forms of energy, commonly expressed in kilowatt hours (kWh) or megawatt hours (MWh).

- **electricity only plant**
A plant designed to produce electricity only.

- **electricity paid by household**
The household paid the electric utility company directly for all household uses of electricity (such as water heating, space heating, air-conditioning, cooking, lighting, and operating appliances.) Bills paid by a third party are not counted as paid by the household.

- **electricity sales**
The amount of kilowatt hours sold in a given period of time; usually grouped by classes of service, such as residential, commercial, industrial, and other. "Other" sales include sales for public street and highway lighting and other sales to public authorities, sales to railroads and railways, and interdepartmental sales.

- **electrochemical process**
The direct process end use in which electricity is used to cause a chemical ransformation. Major uses of electrochemical process occur in the aluminum industry in which alumina is reduced to molten aluminum metal and oxygen, and in the alkalies and chlorine industry, in which brine is separated into caustic soda, chlorine, and hydrogen.

- **elution**
Activities of removing "elutes" a material (uranium) adsorbed on ion exchange resin from the "eluant" solution.

- **emergency backup generation**
The use of electric generators only during interruptions of normal power supply.

- **emergency energy**
Electric energy provided for a limited duration, intended only for use during emergency conditions.

emergency
The failure of an electric power system to generate or deliver electric power as normally intended, resulting in the cutoff or curtailment of service.

emissions
Anthropogenic releases of gases to the atmosphere. In the context of global climate change, they consist of radiatively important greenhouse gases (e.g., the release of carbon dioxide during fuel combustion).

emissions coefficient
A unique value for scaling emissions to activity data in terms of a standard rate of emissions per unit of activity (e.g., pounds of carbon dioxide emitted per Btu of fossil fuel consumed).

end user
A firm or individual that purchases products for its own consumption and not for resale (i.e., an ultimate consumer).

ending stocks
Primary stocks of crude oil and petroleum products held in storage as of 12 midnight on the last day of the month. Primary stocks include crude oil or petroleum products held in storage at (or in) leases, refineries, natural gas processing plants, pipelines, tank farms, and bulk terminals that can store at least 50,000 barrels of petroleum products or that can receive petroleum products by tanker, barge, or pipeline. Crude oil that is in-transit by water from Alaska or that is stored on Federal leases or in the Strategic Petroleum Reserve is included. Primary Stocks exclude stocks of foreign origin that are held in bonded warehouse storage.

energy audit
A program carried out by a utility company in which an auditor inspects a home and suggests ways energy can be saved.

energy broker system
Introduced into Florida by the Public Service Commission, the energy broker system is a system for exchanging information that allows utilities to efficiently exchange hourly quotations of prices at which each is willing to buy and sell electric energy. For the broker system to operate, utility systems must have in place bilateral agreements between all potential parties, must have transmission arrangements between all potential parties, and must have transmission arrangements that allow the exchanges to take place.

energy charge
That portion of the charge for electric service based upon the electric energy (kWh) consumed or billed.

- **energy conservation features**
  This includes building shell conservation features, HVAC conservation features, lighting conservation features, any conservation features, and other conservation features incorporated by the building. However, this category does not include any demand-side management (DSM) program participation by the building. Any DSM program participation is included in the DSM Programs.

- **energy consumption**
  The use of energy as a source of heat or power or as a raw material input to a manufacturing process.

- **energy deliveries**
  Energy generated by one electric utility system and delivered to another system through one or more transmission lines.

- **energy demand**
  The requirement for energy as an input to provide products and/or services.

- **energy effects**
  The changes in aggregate electricity use (measured in megawatt hours) for consumers that participate in a utility DSM (demand-side management) program. Energy effects represent changes at the consumer's meter (i.e., exclude ransmission and distribution effects) and reflect only activities that are undertaken specifically in response to utility-administered programs, including those activities implemented by third parties under contract to the utility. To the extent possible, Energy effects should exclude non-program related effects such as changes in energy usage attributable to non-participants, government-mandated energy-efficiency standards that legislate improvements in building and appliance energy usage, changes in consumer behavior that result in greater energy use after initiation in a DSM program, the natural operations of the marketplace, and weather and business-cycle adjustments.

- **energy efficiency**
  Refers to programs that are aimed at reducing the energy used by specific end-use devices and systems, typically without affecting the services provided. These programs reduce overall electricity consumption (reported in megawatt hours), often without explicit consideration for the timing of program-induced savings. Such savings are generally achieved by substituting technically more advanced equipment to produce the same level of end-use services (e.g. lighting, heating, motor drive) with less electricity. Examples include high-efficiency appliances, efficient lighting

programs, high-efficiency heating, ventilating and air conditioning (HVAC) systems or control modifications, efficient building design, advanced electric motor drives, and heat recovery systems.

- **energy efficient motors**
  Are also known as "high-efficiency motors" and "premium motors." They are virtually interchangeable with standard motors, but differences in construction make them more energy efficient.

- **energy exchange**
  Any transaction in which quantities of energy are received or given up in return for similar energy products.

- **energy expenditures**
  The money directly spent by consumers to purchase energy. Expenditures equal the amount of energy used by the consumer multiplied by the price per unit paid by the consumer.
  Energy from the sun

- **energy information administration (eia)**
  An independent agency within the U.S. Department of Energy that develops surveys, collects energy data, and does analytical and modeling analyses of energy issues. The Agency must satisfy the requests of Congress, other elements within the Department of Energy, Federal Energy Regulatory Commission, the Executive Branch, its own independent needs, and assist the general public, or other interest groups, without taking a policy position.

- **energy information**
  Includes (A) all information in whatever form on fuel reserves, extraction, and energy resources (including petrochemical feed stocks) wherever located; production, distribution, and consumption of energy and fuels wherever carried on; and (B) matters relating to energy and fuels, such as corporate structure and proprietary relationships, costs, prices, capital investment, and assets, and other matters directly related thereto, wherever they exist.

- **energy intensity**
  The ratio of consumption to floor space.

- **energy loss**
  Deleted because there is no need for a general term to encompass all forms of energy loss. Terms referring to losses specific to particular energy sources are defined separately.

- **energy management and control system (emcs)**
  An energy conservation feature that uses mini/microcomputers, instrumentation, control

equipment, and software to manage a building's use of energy for heating, ventilation, air conditioning, lighting, and/or business-related processes. These systems can also manage fire control, safety, and security. Not included as EMCS are time-clock thermostats.

- **energy management practices**
Involvement, as a part of the building's normal operations, in energy efficiency programs that are designed to reduce the energy used by specific end-use systems. This includes the following: EMCS, DSM Program Participation, Energy Audit, and a Building Energy Manager.

- **energy policy act of 1992 (epact)**
This legislation creates a new class of power generators, exempt wholesale generators, that are exempt from the provisions of the Public Holding Company Act of 1935 and grants the authority to the Federal Energy Regulatory Commission to order and condition access by eligible parties to the interconnected transmission grid.

- **energy receipts**
Energy brought into a site from another location.

- **energy reserves**
Estimated quantities of energy sources that are demonstrated to exist with reasonable certainty on the basis of geologic and engineering data (proved reserves) or that can reasonably be expected to exist on the basis of geologic evidence that supports projections from proved reserves (probable/indicated reserves). Knowledge of the location, quantity, and grade of probable/ indicated reserves is generally incomplete or much less certain than it is for proved energy reserves. Note: This term is equivalent to "Demonstrated Reserves" as defined in the resource/reserve classification contained in the U.S. Geological Survey Circular 831, 1980 Demonstrated reserves include measured and indicated reserves but exclude inferred reserves.

- **energy sale(s)**
The transfer of title to an energy commodity from a seller to a buyer for a price or the quantity transferred during a specified period.

- **energy savings**
A reduction in the amount of electricity used by end users as a result of participation in energy efficiency programs and load management programs.

- **energy service provider**
An energy entity that provides service to a retail or end-use customer.

energy source

Any substance or natural phenomenon that can be consumed or transformed to supply heat or power. Examples include petroleum, coal, natural gas, nuclear, biomass, electricity, wind, sunlight, geothermal, water movement, and hydrogen in fuel cells.

energy supplier

Fuel companies supplying electricity, natural gas, fuel oil, kerosene, or LPG (liquefied petroleum gas) to the household.

energy supply

Energy made available for future disposition. Supply can be considered and measured from the point of view of the energy provider or the receiver.

energy

The capacity for doing work as measured by the capability of doing work (potential energy) or the conversion of this capability to motion (kinetic energy). Energy has several forms, some of which are easily convertible and can be changed to another form useful for work. Most of the world's convertible energy comes from fossil fuels that are burned to produce heat that is then used as a transfer medium to mechanical or other means in order to accomplish tasks. Electrical energy is usually measured in kilowatt hours, while heat energy is usually measured in British thermal units (Btu).

energy used in the home

For electricity or natural gas, the quantity is the amount used by the household during the 365- or 366-day period. For fuel oil, kerosene, and liquefied petroleum gas (LPG), the quantity consists of fuel purchased, not fuel consumed. If the level of fuel in the storage tank was the same at the beginning and end of the annual period, then the quantity consumed would be the same as the quantity purchased.

energy-use sectors

A group of major energy-consuming components of U.S. society developed to measure and analyze energy use. The sectors most commonly referred to in EIA are: residential, commercial, industrial, transportation, and electric power.

energy-weighted industrial output

The weighted sum of real output for all two-digit Standard Industrial Classification (SIC) manufacturing industries plus agriculture, construction, and mining. The weight for each industry is the ratio between the quantity of end-use energy consumption to the value of real output.

engine size

The total volume within all cylinders of an engine when

pistons are at their lowest positions. The engine is usually measured in "liters" or "cubic inches of displacement (CID)." Generally, larger engines result in greater engine power, but less fuel efficiency. There are 61.024 cubic inches in a liter.

- **enriched uranium**

  Uranium in which the U-235 isotope concentration has been increased to greater than the 0.711 percent U-235 (by weight) present in natural uranium.

- **enrichment feed deliveries**

  Uranium that is shipped under contract to a supplier of enrichment services for use in preparing enriched uranium product to a specified U-235 concentration and that ultimately will be used as fuel in a nuclear reactor.

- **enrichment tails assay**

  A measure of the amount of fissile uranium (U-235) remaining in the waste stream from the uranium enrichment process. The natural uranium "feed" that enters the enrichment process generally contains 0.711 percent (by weight) U-235. The "product stream" contains enriched uranium (more than 0.711 percent U-235) and the "waste" or "tails" stream contains depleted uranium (less than 0.711 percent U-235). At the historical enrichment tails assay of 0.2 percent, the waste stream would contain 0.2 percent U-235. A higher enrichment tails assay requires more uranium feed (thus permitting natural uranium stockpiles to be decreased), while increasing the output of enriched material for the same energy expenditure.

- **environmental impact statement**

  A report that documents the information required to evaluate the environmental impact of a project. It informs decision makers and the public of the reasonable alternatives that would avoid or minimize adverse impacts or enhance the quality of the environment.

- **environmental protection agency (epa) certification files**

  Computer files produced by EPA for analysis purposes. For each vehicle make, model and year, the files contain the EPA test MPGs (city, highway, and 55/45 composite). These MPG's are associated with various combinations of engine and drive-train technologies (e.g., number of cylinders, engine size, gasoline or diesel fuel, and automatic or manual transmission). These files also contain information similar to that in the DOE/EPA Gas Mileage Guide, although the MPGs in that publication are adjusted for shortfall.

- **environmental restoration**
  Although usually described as "cleanup," this function encompasses a wide range of activities, such as stabilizing contaminated soil; treating ground water; decommissioning process buildings, nuclear reactors, chemical separations plants, and many other facilities; and exhuming sludge and buried drums of waste.

- **environmental restrictions**
  In reference to coal accessibility, land-use restrictions that constrain, postpone, or prohibit mining in order to protect environmental resources of an area; for example, surface- or groundwater quality, air quality affected by mining, or plants or animals or their habitats.

- **epa certification**
  A permanent label on fireplace inserts and freestanding wood stoves manufactured after July 1, 1988, indicating that the equipment meets EPA standards for clean burning.

- **epa composite mpg**
  The harmonic mean of the EPA city and highway MPG (miles per gallon), weighted under the assumption of 55 percent city driving and 45 percent highway driving.

- **equilibrium cycle**
  An analytical term that refers to fuel cycles that occur after the initial one or two cycles of a reactor's operation. For a given type of reactor, equilibrium cycles have similar fuel characteristics.

- **equity (financial)**
  Ownership of shareholders in a corporation represented by stock.

- **equity capital**
  The sum of capital from retained earnings and the issuance of stock.

- **equity crude oil**
  The proportion of production that a concession owner has the legal and contractual right to retain.

- **equity in earnings of unconsolidated affiliates**
  A company's proportional share (based on ownership) of the net earnings or losses of an unconsolidated affiliate.

- **establishment**
  An economic unit, generally, at a single physical location where business is conducted or where services or industrial operations are performed. However, "establishment" is not synonymous with "building."

- **estimated additional resources (ear)**
  The uranium in addition to reasonable assured resources (RAR) that is expected to occur, mostly on the basis of direct geological evidence, in extensions of well-explored deposits, little-

explored deposits, and undiscovered deposits believed to exist along a well-defined geologic trend with known deposits, such that the uranium can subsequently be recovered within the given cost ranges. Estimates of tonnage and grade are based on available sampling data and on knowledge of the deposit characteristics as determined in the best known parts of the deposit or in similar deposits. EAR correspond to DOE's Probable Potential Resource Category.

- **etbe (ethyl tertiary butyl ether)(CH3)3COC2H**

An oxygenate blend stock formed by the catalytic etherification of isobutylene with ethanol.

- **ethane (c2h6)**

A normally gaseous straight-chain hydrocarbon. It is a colorless paraffinic gas that boils at a temperature of -127.48 degrees Fahrenheit. It is extracted from natural gas and refinery gas streams.

- **ethanol (ch3-ch2oh)**

A clear, colorless, flammable oxygenated hydrocarbon. Ethanol is typically produced chemically from ethylene, or biologically from fermentation of various sugars from carbohydrates found in agricultural crops and cellulosic residues from crops or wood. It is used in the United States as a gasoline octane enhancer and oxygenate (blended up to 10 percent concentration). Ethanol can also be used in high concentrations (E85) in vehicles designed for its use. The lower heating value, equal to 76,000 Btu per gallon, is assumed for estimates in the Renewables Energy Annual report.

- **ether**

A generic term applied to a group of organic chemical compounds composed of carbon, hydrogen, and oxygen, characterized by an oxygen atom attached to two carbon atoms (e.g., methyl tertiary butyl ether).

- **ethylene**

An olefinic hydrocarbon recovered from refinery processes or petrochemical processes. Ethylene is used as a petrochemical feedstock for numerous chemical applications and the production of consumer goods.

- **ethylene dichloride**

A colorless, oily liquid used as a solvent and fumigant for organic synthesis, and for ore flotation.

- **evacuated-tube collector**

A collector in which solar thermal heat is captured by use of a collector fluid that flows through an absorber tube contained inside an evacuated glass tube.

- **evaporation pond**

A containment pond (that preferably has an impermeable

lining of clay or synthetic material such as hypalon) to hold liquid wastes and to concentrate the waste through evaporation.

- **evaporative cooler (swamp cooler)**
  An air-cooling unit that turns air into moist, cool air by saturating the air with water vapor. It does not cool air by use of a refrigeration unit.

- **excess statutory depletion**
  The excess of estimated statutory depletion allowable as an income tax deduction over the amount of cost depletion otherwise allowable as a tax deduction, determined on a total enterprise basis.

- **exchange agreement**
  A contractual agreement in which quantities of crude oil, petroleum products, natural gas, or electricity are delivered, either directly or through intermediaries, from one company to another company, in exchange for the delivery by the second company to the first company of an equivalent volume or heat content. The exchange may take place at the same time and location or at different times and/or locations. Such agreements may also involve the payment of cash.

- **exchange, electricity**
  A type of energy exchange in which one electric utility agrees to supply electricity to another. Electricity received is returned in kind at a later time or is accumulated as an energy balance until the end of a specified period, after which settlement may be made by monetary payment. Note: This term is also referred to as exchange energy.

- **exchange, natural gas**
  A type of energy exchange in which one company agrees to deliver gas, either directly or through intermediaries, to another company at one location or in one time period in exchange for the delivery by the second company to the first company of an equivalent volume or heat content at a different location or time period.

- **exchange, petroleum**
  A type of energy exchange in which quantities of crude oil or any petroleum product(s) are received or given up in return for other crude oil or petroleum products. It includes reciprocal sales and purchases.

- **exchange, power**
  Delete in favor of the already-defined term **exchange energy**, which should be renamed **exchange electricity** or **exchange, electricity**.

- **exempt wholesale generator**
  Wholesale generators created under the 1992 Energy Policy Act

that are exempt from certain financial and legal restrictions stipulated in the Public Utilities Holding Company Act of 1935.

- **exhaust fan**
Small fans located in the wall or ceiling that exhaust air, odors, and moisture from the bathroom, kitchen, or basement to the outside.

- **expenditure**
The incurrence of a liability to obtain an asset or service.

- **expenditures per million btu**
The aggregate ratio of a group of buildings' total expenditures for a given fuel to the total consumption of that fuel.

- **expenditures per square foot**
The aggregate ratio of a group of buildings' total expenditures for a given fuel to the total floor space in those buildings.

- **exploration drilling**
Drilling done in search of new mineral deposits, on extensions of known ore deposits, or at the location of a discovery up to the time when the company decides that sufficient ore reserves are present to justify commercial exploration. Assessment drilling is reported as exploration drilling.

- **exploratory well**
A hole drilled: a) to find and produce oil or gas in an area previously considered unproductive area; b) to find a new reservoir in a known field, i.e., one previously producing oil and gas from another reservoir, or c) to extend the limit of a known oil or gas reservoir.

- **exports**
Shipments of goods from within the 50 States and the District of Columbia to U.S. possessions and territories or to foreign countries.

- **extensions**
Any new reserves credited to a previously producing reservoir because of enlargement of its proved area. This enlargement in proved area is usually due to new well drilling outside of the previously known productive limits of the reservoir.

- **extensions, discoveries, and other additions**
Additions to an enterprise's proved reserves that result from (1) extension of the proved acreage of previously discovered (old) reserves through additional drilling in periods subsequent to discovery and (2) discovery of new fields with proved reserves or of new reservoirs of proved reserves in old fields.

- **externalities**
Benefits or costs, generated as a byproduct of an economic activity, that do not accrue to the parties involved in the activity. Environmental externalities are

benefits or costs that manifest themselves through changes in the physical or biological environment.

- **extraction loss**

The reduction in volume of natural gas due to the removal of natural gas liquid constituents such as ethane, propane, and butane at natural gas processing plants.

- **extractive industries**

Industries involved in the activities of (1) prospecting and exploring for wasting (non-regenerative) natural resources, (2) acquiring them, (3) further exploring them, (4) developing them, and (5) producing (extracting) them from the earth. The term does not encompass the industries of forestry, fishing, agriculture, animal husbandry, or any others that might be involved with resources of a regenerative nature.

- **extraordinary income deductions (electric utility)**

Those items related to transactions of a nonrecurring nature that are not typical or customary business activities of the utility and that would significantly distort the current year's net income if reported other than as extraordinary items.

- **f.a.s. value**

Free alongside ship value. The value of a commodity at the port of exportation, generally including the purchase price plus all charges incurred in placing the commodity alongside the carrier at the port of exportation in the country of exportation.

- **f.o.b. price**

The price actually charged at the producing country's port of loading. The reported price should be after deducting any rebates and discounts or adding premiums where applicable and should be the actual price paid with no adjustment for credit terms.

- **f.o.b. value (coal)**

Free-on-board value. This is the value of coal at the coal mine or of coke and breeze at the coke plant without any insurance or freight transportation charges added.

- **fabricated fuel**

Fuel assemblies composed of an array of fuel rods loaded with pellets of enriched uranium dioxide.

- **facilities charge**

An amount to be paid by the customer in a lump sum, or periodically as reimbursement for facilities furnished. The charge may include operation and maintenance as well as fixed costs.

facility

An existing or planned location or site at which prime movers, electric generators, and/or equipment for converting mechanical, chemical, and/or nuclear energy into electric energy are situated or will be situated. A facility may contain more than one generator of either the same or different prime mover type. For a cogenerator, the facility includes the industrial or commercial process.

fahrenheit

A temperature scale on which the boiling point of water is at 212 degrees above zero on the scale and the freezing point is at 32 degrees above zero at standard atmospheric pressure.

failure or hazard

Any electric power supply equipment or facility failure or other event that, in the judgment of the reporting entity, constitutes a hazard to maintaining the continuity of the bulk electric power supply system such that a load reduction action may become necessary and reportable outage may occur. Types of abnormal conditions that should be reported include the imposition of a special operating procedure, the extended purchase of emergency power, other bulk power system actions that may be caused by a natural disaster, a major equipment failure that would impact the bulk power supply, and an environmental and/or regulatory action requiring equipment outages.

farm out (in) arrangement

An arrangement, used primarily in the oil and gas industry, in which the owner or lessee of mineral rights (the first party) assigns a working interest to an operator (the second party), the consideration for which is specified exploration and/or development activities. The first party retains an overriding royalty or other type of economic interest in the mineral production. The arrangement from the viewpoint of the second party is termed a "farm-in arrangement."

farm use

Energy use at establishments where the primary activity is growing crops and/or raising animals. Energy use by all facilities and equipment at these establishments is included, whether or not it is directly associated with growing crops and/or raising animals. Common types of energy-using equipment include tractors, irrigation pumps, crop dryers, smudge pots, and milking machines. Facility energy use encompasses all structures at the establishment, including the farm house.

fast breeder reactor (fbr)

A reactor in which the fission chain reaction is sustained primarily by fast neutrons rather than by thermal or intermediate neutrons. Fast reactors require little or no moderator to slow down the neutrons from the speeds at which they are ejected from fissioning nuclei. This type of reactor produces more fissile material than it consumes.

federal coal lease

A lease granted to a mining company to produce coal from land owned and administered by the Federal Government in exchange for royalties and other revenues.

federal electric utility

A utility that is either owned or financed by the Federal Government.

federal energy regulatory commission (ferc)

The Federal agency with jurisdiction over interstate electricity sales, wholesale electric rates, hydroelectric licensing, natural gas pricing, oil pipeline rates, and gas pipeline certification. FERC is an independent regulatory agency within the Department of Energy and is the successor to the Federal Power Commission.

federal power act

Enacted in 1920, and amended in 1935, the Act consists of three parts. The first part incorporated the Federal Water Power Act administered by the former Federal Power Commission, whose activities were confined almost entirely to licensing non-Federal hydroelectric projects. Parts II and III were added with the passage of the Public Utility Act. These parts extended the Act's jurisdiction to include regulating the interstate transmission of electrical energy and rates for its sale as wholesale in interstate commerce. The Federal Energy Regulatory Commission is now charged with the administration of this law.

federal power commission (fpc)

The predecessor agency of the Federal Energy Regulatory Commission. The Federal Power Commission was created by an Act of Congress under the Federal Water Power Act on June 10, 1920. It was charged originally with regulating the electric power and natural gas industries. It was abolished on September 30, 1977, when the Department of Energy was created. Its functions were divided between the Department of Energy and the Federal Energy Regulatory Commission, an independent regulatory agency.

federal region

In a Presidential directive issued in 1969, various Federal agencies (among them the currently designated Department of Health and Human Services, the Department of Labor, the Office of Economic Opportunity, and the Small Business Administration) were instructed to adopt a uniform field system of 10 geographic regions with common boundaries and headquarters cities. The action was taken to correct the evolution of fragmented Federal field organization structures that each agency or component created independently, usually with little reference to other agencies' arrangements. Most Federal domestic agencies or their components have completed realignments and relocations to conform to the Standard Federal Administration Regions (SFARs).

- **fee interest**

The absolute, legal possession and ownership of land, property, or rights, including mineral rights. A fee interest can be sold (in its entirety or in part) or passed on to heirs or successors.

- **feeder line**

An electrical line that extends radially from a distribution substation to supply electrical energy within an electric area or sub-area.

- **ferc guidelines**

A compilation of the Federal Energy Regulatory Commission's enabling statutes; procedural and program regulations; and orders, opinions, and decisions.

- **ferc**

The Federal Energy Regulatory Commission.

- **fertile material**

Material that is not itself fissionable by thermal neutrons but can be converted to fissile material by irradiation. The two principal fertile materials are uranium-238 and thorium-232.

- **field**

An area consisting of a single reservoir or multiple reservoirs all grouped on, or related to, the same individual geological structural feature and/or stratigraphic condition. There may be two or more reservoirs in a field that are separated vertically by intervening impervious strata or laterally by local geologic barriers, or by both.

- **field area**

A geographic area encompassing two or more pools that have a common gathering and metering system, the reserves of which are reported as a single unit. This concept applies primarily to the Appalachian region.

- **field discovery year**

The calendar year in which a field was first recognized as containing economically recoverable accumulations of oil and/or gas.

field production

Represents crude oil production on leases, natural gas liquids production at natural gas processing plants, new supply of other hydrocarbons/oxygenates and motor gasoline blending components, and fuel ethanol blended into finished motor gasoline.

field separation facility

A surface installation designed to recover lease condensate from a produced natural gas stream usually originating from more than one lease and managed by the operator of one or more of these leases.

file rate schedule

The rate for a particular electric service, including attendant contract terms and conditions, accepted for filing by a regulatory body with appropriate oversight authority.

filing

Any written application, complaint, declaration, petition, protest, answer, motion, brief, exception, rate schedule, or other pleading, amendment to a pleading, document, or similar paper that is submitted to a utility commission.

final order

A final ruling by FERC that terminates an action, decides some matter litigated by the petitioning parties, operates to some right, or completely disposes of the subject matter.

financial accounting standards board (fasb)

An independent board responsible, since 1973, for establishing generally accepted accounting principles. Its official pronouncement are called "Statements of Financial Accounting Standards" and "Interpretations of Financial Accounting Standards."

finished leaded gasoline

Contains more than 0.05 gram of lead per gallon or more than 0.005 gram of phosphorus per gallon. Premium and regular grades are included, depending on the octane rating. Includes leaded gasohol. Blendstock is excluded until blending has been completed. Alcohol that is to be used in the blending of gasohol is also excluded.

finished motor gasoline

A complex mixture of relatively volatile hydrocarbons, with or without small quantities of additives, blended to form a fuel suitable for use in spark-ignition engines. Specification for motor gasoline, as given in ASTM Specification D439-88 or Federal Specification VV-G-1690B, include a boiling range of 122 degrees to 158 degrees Fahrenheit at the 10-

percent point to 365 degrees to 374 degrees Fahrenheit at the 90-percent point and a Reid vapor pressure range from 9 to 15 psi. "Motor gasoline" includes finished leaded gasoline, finished unleaded gasoline, and gasohol. Blendstock is excluded until blending has been completed. (Alcohol that is to be used in the blending of gasohol is also excluded.)

▪ **finished unleaded gasoline**

Contains not more than 0.05 gram of lead per gallon and not more than 0.005 gram of phosphorus per gallon. Premium and regular grades are included, depending on the octane rating. Includes unleaded gasohol. Blendstock is excluded until blending has been completed. Alcohol that is to be used in the blending of gasohol is also excluded.

▪ **firm**

An association, company, corporation, estate, individual, joint venture, partnership, or sole proprietorship, or any other entity, however organized, including: (a) charitable or educational institutions; (b) the Federal Government, including corporations, departments, Federal agencies, and other instrumentalities; and State and local governments. A firm may consist of (1) a parent entity, including the consolidated and unconsolidated entities (if any) that it directly or indirectly controls; (2) a parent and its consolidated entities only; (3) an unconsolidated entity; or (4) any part or combination of the above.

▪ **firm power**

Power or power-producing capacity, intended to be available at all times during the period covered by a guaranteed commitment to deliver, even under adverse conditions.

▪ **first purchase (of crude oil)**

An equity (not custody) transaction commonly associated with a transfer of ownership of crude oil associated with the physical removal of the crude oil from a property for the first time (also referred to as a lease sale). A first purchase normally occurs at the time and place of ownership transfer where the crude oil volume sold is measured and recorded on a run ticket or other similar physical evidence of purchase. The volume purchased and the cost of such transaction shall not be measured farther from the wellhead than the point at which the value for landowner royalties is established, if there was a separate landowner.

▪ **first purchase price**

The marketed first sales price of domestic crude oil, consistent with the removal price defined by the provisions of the Windfall Profits Tax on Domestic Crude Oil.

■ **fiscal year**

The U.S. Government's fiscal year runs from October 1 through September 30. The fiscal year is designated by the calendar year in which it ends; e.g., fiscal year 2002 begins on October 1, 2001 and ends on September 30, 2002

■ **fissile material**

Material that can be caused to undergo atomic fission when bombarded by neutrons. The most important fissionable materials are uranium-235, plutonium-239, and uranium-233.

■ **fission**

The process whereby an atomic nucleus of appropriate type, after capturing a neutron, splits into (generally) two nuclei of lighter elements, with the release of substantial amounts of energy and two or more neutrons.

■ **fixed asset turnover**

A ratio of revenue to fixed assets which is a measure of the productivity and efficiency of property, plant, and equipment in generating revenue. A high turnover reflects positively on the entity's ability to utilize properly its fixed assets in business operations.

■ **fixed assets**

Tangible property used in the operations of an entity, but not expected to be consumed or converted into cash in the ordinary course of events. With a life in excess of one year, not intended for resale to customers, and subject to depreciation (with the exception of land), they are usually referred to as property, plant, and equipment.

■ **fixed carbon**

The nonvolatile matter in coal minus the ash. Fixed carbon is the solid residue other than ash obtained by prescribed methods of destructive distillation of a coal. Fixed carbon is the part of the total carbon that remains when coal is heated in a closed vessel until all matter is driven off.

■ **fixed charge coverage**

The ratio of earnings available to pay so-called fixed charges to such fixed charges. Fixed charges include interest on funded debt, including leases, plus the related amortizations of debt discount, premium, and expense. Earnings available for fixed charges may be computed before or after deducting income taxes. Occasionally credits for the "allowance for funds used during construction" are excluded from the earnings figures. The precise procedures followed in calculating fixed charge or interest coverage vary widely.

■ **fixed cost (expense)**

An expenditure or expense that does not vary with volume level of activity.

■ **fixed operating costs**

Costs other than those associated with capital investment that do not vary with the operation, such as maintenance and payroll.

■ **flared**

Gas disposed of by burning in flares usually at the production sites or at gas processing plants.

■ **flat and meter rate schedule**

An electric rate schedule consisting of two components, the first of which is a service charge and the second a price for the energy consumed.

■ **flat demand rate schedule**

An electric rate schedule based on billing demand that provides no charge for energy.

■ **flat plate pumped**

A medium-temperature solar thermal collector that typically consists of a metal frame, glazing, absorbers (usually metal), and insulation and that uses a pumped liquid as the heat-transfer medium: predominant use is in water-heating applications.

■ **fleet vehicle**

Any motor vehicle a company owns or leases that is in the normal operations of a company. Vehicles which are used in the normal operation of a company, but are owned by company employees are not fleet vehicles. If a company provides services in addition to providing natural gas, only those vehicles that are used by the natural gas provider portion of a company should be counted as fleet vehicles. Vehicles that are considered "off-road" (e.g., farm or construction vehicles) or demonstration vehicles are not to be counted as fleet vehicles. Fleet vehicles include gasoline/diesel powered vehicles and alternative-fuel vehicles.

■ **flexible fuel vehicle**

A vehicle that can operate on (1) alternative fuels (such as M85 or E85) (2) 100-percent petroleum-based fuels (3) any mixture of an alternative fuel (or fuels) and a petroleum-based fuel. Flexible fuel vehicles have a single fuel system to handle alternative and petroleum-based fuels. Flexible fuel vehicle and variable fuel vehicle are synonymous terms.

■ **flexicoking**

A thermal cracking process which converts heavy hydrocarbons such as crude oil, tar sands bitumen, and distillation residues into light hydrocarbons. Feed stocks can be any pumpable hydrocarbons, including those containing high concentrations of sulfur and metals.

■ **flexicoking**

A thermal cracking process which converts heavy hydrocarbons such as crude oil, tar sands bitumen, and distillation residues into light hydrocarbons. Feed stocks can be any pumpable hydrocarbons including those containing high concentrations of sulfur and metals.

- **floor (coal)**

The upper surface of the stratum underlying a coal seam. In coals that were formed in persistent swamp environments, the floor is typically a bed of clay, known as "underclay," representing the soil in which the trees or other coal-forming swamp vegetation was rooted.

- **floor price**

A price specified in a market-price contract as the lowest purchase price of the uranium, even if the market price falls below the specified price. The floor price may be related to the seller's production costs.

- **floor space**

The area enclosed by exterior walls of a building, including parking areas, basements, or other floors below ground level. It is measured in square feet.

- **floor, wall, or pipeless furnace**

Space-heating equipment consisting of a ductless combustor or resistance unit, having an enclosed chamber where fuel is burned or where electrical-resistance heat is generated to warm the rooms of a building. A floor furnace is located below the floor and delivers heated air to the room immediately above or (if under a partition) to the room on each side. A wall furnace is installed in a partition or in an outside wall and delivers heated air to the rooms on one or both sides of the wall. A pipeless furnace is installed in a basement and delivers heated air through a large register in the floor of the room or hallway immediately above.

- **flow control**

The laws, regulations, and economic incentives or disincentives used by waste managers to direct waste generated in a specific geographic area to a designated landfill, recycling, or waste-to-energy facility.

- **flue**

An enclosed passageway for directing products of combustion to the atmosphere.

- **flue gas desulfurization**

Equipment used to remove sulfur oxides from the combustion gases of a boiler plant before discharge to the atmosphere. Also referred to as scrubbers. Chemicals such as lime are used as scrubbing media.

- **flue-gas desulfurization unit (scrubber)**

Equipment used to remove sulfur oxides from the combustion gases of a boiler plant before discharge to the atmosphere. Chemicals such as lime are used as the scrubbing media.

- **flue-gas particulate collector**

Equipment used to remove fly ash from the combustion gases of a boiler plant before discharge to the atmosphere. Particulate collectors include electrostatic precipitators, mechanical collectors (cyclones), fabric filters (baghouses), and wet scrubbers.

- **fluid catalytic cracking**

The refining process of breaking down the larger, heavier, and more complex hydrocarbon molecules into simpler and lighter molecules. Catalytic cracking is accomplished by the use of a catalytic agent and is an effective process for increasing the yield of gasoline from crude oil. Catalytic cracking processes fresh feeds and recycled feeds.

- **fluid coking**

A thermal cracking process utilizing the fluidized-solids technique to remove carbon (coke) for continuous conversion of heavy, low-grade oils into lighter products.

- **fluid coking**

A thermal cracking process utilizing the fluidized-solids technique to remove carbon (coke) for continuous conversion of heavy, low-grade oils into lighter products.

- **fluidized-bed combustion**

A method of burning particulate fuel, such as coal, in which the amount of air required for combustion far exceeds that found in conventional burners. The fuel particles are continually fed into a bed of mineral ash in the proportions of 1 part fuel to 200 parts ash, while a flow of air passes up through the bed, causing it to act like a turbulent fluid.

- **fluorescent lamp**

A glass enclosure in which light is produced when electricity is passed through mercury vapor inside the enclosure. The electricity creates a radiation discharge that strikes a coating on the inside surface of the enclosure, causing the coating to glow.

- **fluorescent light bulbs**

These are usually long, narrow, white tubes made of glass coated on the inside with fluorescent material, which is connected to a fixture at both ends of the light bulb; some are circular tubes. The light bulb produces light by passing electricity through mercury vapor, which causes the fluorescent coating to glow or fluoresce.

- **fluorescent lighting other than compact fluorescent bulbs**

In fluorescent lamps, energy is converted to light by using an electric charge to "excite" gaseous atoms within a fluorescent tube. Common types are "cool white," "warm white," etc. Special energy efficient fluorescent lights have been developed that produce the same amount of light while consuming less energy.

- **flux material**

A substance used to promote fusion, e.g., of metals or minerals.

- **fly ash**

Particulate matter mainly from coal ash in which the particle diameter is less than 1×10^4 meter. This ash is removed from the flue gas using flue gas particulate collectors such as fabric filters and electrostatic precipitators.

- **fme**

Free Market Economies. Countries that are members of the Council for Mutual Economic Assistance (CMEA) are not included.

- **footage drilled**

Total footage for wells in various categories, as reported for any specified period, includes (1) the deepest total depth (length of well bores) of all wells drilled from the surface, (2) the total of all bypassed footage drilled in connection with reported wells, and (3) all new footage drilled for directional sidetrack wells. Footage reported for directional sidetrack wells does not include footage in the common bore that is reported as footage for the original well. In the case of old wells drilled deeper, the reported footage is that which was drilled below the total depth of the old well.

- **forced outage**

The shutdown of a generating unit, transmission line, or other facility for emergency reasons or a condition in which the generating equipment is unavailable for load due to unanticipated breakdown.

- **foreign access**

Refers to proved reserves of crude, condensate, and natural gas liquids applicable to long-term supply agreements with foreign governments or authorities in which the company or one of its affiliates acts as producer.

- **foreign currency transaction gains and losses**

Gains or losses resulting from the effect of exchange rate changes on transactions denominated in currencies other than the functional currency (for example, a U.S. enterprise may borrow Swiss francs or a French subsidiary may have a receivable denominated in kroner from a Danish customer). Gains and losses on those foreign currency transactions are generally included in determining net income for the period in which exchange rates change unless the transaction hedges a foreign currency commitment or a net investment in a foreign entity. Inter company transactions of a long-term investment nature are considered part of a parent's net investment and hence do not give rise to gains or losses.

- **foreign currency translation effects**
 Gains or losses resulting from the process of expressing amounts denominated or measured in one currency in terms of another currency by use of the exchange rate between the two currencies. This process is generally required to consolidate the financial statements of foreign affiliates into the total company financial statements and to recognize the conversion of foreign currency or the settlement of a receivable or payable denominated in foreign currency at a rate different from that at which the item is recorded. Translation adjustments are not included in determining net income, but are disclosed as separate components of consolidated equity.

- **foreign operations**
 These are operations that are located outside the United States. Determination of whether an enterprise's mobile assets, such as offshore drilling rigs or ocean-going vessels, constitute foreign operations should depend on whether such assets are normally identified with operations located outside the United States.

- **foreign-controlled firms (coal)**
 Foreign-controlled firms are U.S. coal producers with more than 50 percent of their stock or assets owned by a foreign firm.

- **forward cost (1)**
 Forward costs are those operating and capital costs yet to be incurred at the time an estimate of reserves is made. Profits and "sunk" costs, such as past expenditures for property acquisition, exploration, and mine development, are not included. Therefore, the various forward-cost categories are independent of the market price at which uranium produced from the reserves would be sold.

- **forward cost (2)**
 The operating and capital costs still to be incurred in the production of uranium from in-place reserves. By using forward costing, estimates for reserves for ore deposits in differing geological settings and status of development can be aggregated and reported for selected cost categories. Included are costs for labor, materials, power and fuel, royalties, payroll taxes, insurance, and applicable general and administrative costs. Excluded from forward cost estimates are prior expenditures, if any, incurred

for property acquisition, exploration, mine development, and mill construction, as well as income taxes, profit, and the cost of money. Forward costs are neither the full costs of production nor the market price at which the uranium, when produced, might be sold.

- **forward costs (uranium)**
  The operating and capital costs that will be incurred in any future production of uranium from in-place reserves. Included are costs for labor, materials, power and fuel, royalties, payroll taxes, insurance, and general and administrative costs that are dependent upon the quantity of production and, thus, applicable as variable costs of production. Excluded from forward costs are prior expenditures, if any, incurred for property acquisition, exploration, mine development, and mill construction, as well as income taxes, profit, and the cost of money. Note: By use of forward costing, estimates of reserves for ore deposits in differing geological settings can be aggregated and reported as the maximum amount that can theoretically be extracted to recover the specified costs of uranium oxide production under the listed forward cost categories.

- **forward coverage**
  Amount of uranium required to assure uninterrupted operation of nuclear power plants.

- **fossil fuel**
  An energy source formed in the earths crust from decayed organic material. The common fossil fuels are petroleum, coal, and natural gas.

- **fossil fuel depletion**

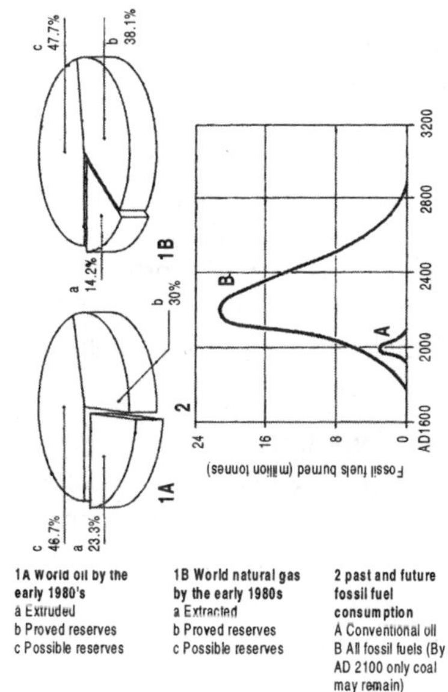

1A World oil by the early 1980's
a Extruded
b Proved reserves
c Possible reserves

1B World natural gas by the early 1980s
a Extracted
b Proved reserves
c Possible reserves

2 past and future fossil fuel consumption
A Conventional oil
B All fossil fuels (By AD 2100 only coal may remain)

- **fossil fuel plant**
  A plant using coal, petroleum, or gas as its source of energy.

- **fossil fuel steam-electric power plant**
  An electricity generation plant in which the prime mover is a turbine rotated by high-pressure

steam produced in a boiler by heat from burning fossil fuels.

- **fossil-fuel electric generation**

Electric generation in which the prime mover is a turbine rotated by high-pressure steam produced in a boiler by heat from burning fossil fuels.

- **foundry**

An operation where metal castings are produced, using coke as a fuel.

- **foundry coke**

This is a special coke that is used in furnaces to produce cast and ductile iron products. It is a source of heat and also helps maintain the required carbon content of the metal product. Foundry coke production requires lower temperatures and longer times than blast furnace coke.

- **fractionation**

The process by which saturated hydrocarbons are removed from natural gas and separated into distinct products, or "fractions," such as propane, butane, and ethane.

- **framework convention on climate change (fccc)**

An agreement opened for signature at the "Earth Summit" in Rio de Janeiro, Brazil, on June 4, 1992, which has the goal of stabilizing greenhouse gas concentrations in the atmosphere at a level that would prevent significant anthropogenically forced climate change.

- **free alongside ship (f.a.s.)**

The value of a commodity at the port of exportation, generally including the purchase price plus all charges incurred in placing the commodity alongside the carrier at the port of exportation.

- **free on board (f.o.b.)**

A sales transaction in which the seller makes the product available for pick up at a specified port or terminal at a specified price and the buyer pays for the subsequent transportation and insurance.

- **free on board (f.o.b.)**

A sales transaction in which the seller makes the product available for pick up at a specified port or terminal at a specified price and the buyer pays for the subsequent transportation and insurance.

- **free well**

A well drilled and equipped by an assignee as consideration for the assignment of a fractional share of the working interest, commonly under a farm-out agreement.

- **fresh feed input**

Represents input of material (crude oil, unfinished oils, natural gas liquids, other hydrocarbons and oxygenates or finished products) to processing units at a refinery that is being processed (input) into a particular unit for the first time. Examples: 1. unfinished oils coming out of a crude oil distillation unit that are put into a catalytic cracking unit are considered fresh feed to the catalytic cracking unit. 2. unfinished oils coming out of a catalytic cracking unit being looped back into the same catalytic cracking unit to be reprocessed are not considered fresh feed.

- **fresh feeds**

Crude oil or petroleum distillates that are being fed to processing units for the first time.

- **fuel**

Any material substance that can be consumed to supply heat or power. Included are petroleum, coal, and natural gas (the fossil fuels), and other consumable materials, such as uranium, biomass, and hydrogen.

- **fuel cell**

A device capable of generating an electrical current by converting the chemical energy of a fuel (e.g., **hydrogen**) directly into electrical energy. Fuel cells differ from conventional electrical cells in that the active materials such as fuel and oxygen are not contained within the cell but are supplied from outside. It does not contain an intermediate heat cycle, as do most other electrical generation techniques.

- **fuel cycle**

The entire set of sequential processes or stages involved in the utilization of fuel, including extraction, transformation, transportation, and combustion. Emissions generally occur at each stage of the fuel cycle.

- **fuel emergencies**

An emergency that exists when supplies of fuels or hydroelectric storage for generation are at a level or estimated to be at a level that would threaten the reliability or adequacy of bulk electric power supply. The following factors should be taken into account to determine that a fuel emergency exists:

  - Fuel stock or hydroelectric project water storage levels are 50 percent or less of normal for that particular time of the year and a continued downward trend in fuel stock or hydroelectric project water storage level is estimated; or
  - Unscheduled dispatch or emergency generation is causing an abnormal use of a particular fuel type, such that the future supply of stocks of that fuel could reach a level that threatens the reliability or adequacy of bulk electric power supply.

- **fuel ethanol (ch3 – ch2oh)**

An anhydrous denatured aliphatic alcohol intended for gasoline blending as described in the definition of oxygenates. It is also used in high concentrations to produce E85.

- **fuel expenses**

These costs include the fuel used in the production of steam or driving another prime mover for the generation of electricity. Other associated expenses include unloading the shipped fuel and all handling of the fuel up to the point where it enters the first bunker, hopper, bucket, tank, or holder in the boiler-house structure.

- **fuel injection**

A fuel delivery system whereby gasoline is pumped to one or more fuel injectors under high pressure. The fuel injectors are valves that, at the appropriate times, open to allow fuel to be sprayed or atomized into a throttle bore or into the intake manifold ports. The fuel injectors are usually solenoid operated valves under the control of the vehicle's on-board computer (thus the term "electronic fuel injection"). The fuel efficiency of fuel injection systems is less temperature-dependent than carburetor systems. Diesel engines always use injectors.

- **fuel oil**

A liquid petroleum product less volatile than gasoline, used as an energy source. Fuel oil includes distillate fuel oil (No. 1, No. 2, and No. 4), and residual fuel oil (No. 5 and No. 6).

- **fuel purchase agreement**

An agreement between a company and a fuel provider which stipulates that the company agrees to purchase its fuel from the fuel provider. If the company has a credit card for use at a fuel provider's locations, but is not bound by an additional agreement to purchase fuel from that provider, the credit card agreement alone is not considered a fuel purchase agreement.

- **fuel ratio**

The ratio of fixed carbon to volatile matter in coal.

- **fuel switching capability**

The short-term capability of a manufacturing establishment to have used substitute energy sources in place of those actually consumed. Capability to use substitute energy sources means that the establishment's combustors (for example, boilers, furnaces, ovens, and blast furnaces) had the machinery or equipment either in place or available for installation so that substitutions could actually have been introduced within 30 days without extensive modifications. Fuel-switching capability does not depend on the relative prices of energy sources; it depends only on the characteristics of the equipment and certain legal constraints.

- **fuel wood**
  Wood and wood products, possibly including scrubs and branches, etc, bought or gathered, and used by direct combustion.

- **fuel/fabricator assembly identifier**
  Individual assembly identifier based on a numbering scheme developed by individual fuel fabricators. Most fuel fabricator assembly identifiers schemes closely match the scheme developed by the American National Standards Institute (ANSI) and are therefore unique.

- **fuels solvent deasphalting**
  A refining process for removing asphalt compounds from petroleum fractions, such as reduced crude oil. The recovered stream from this process is used to produce fuel products.

- **fuel-switching dsm program assistance**
  DSM program assistance where the sponsor encourages consumers to change from one fuel to another for a particular end-use service. For example, utilities might encourage consumers to replace electric water heaters with gas units or encourage industrial consumers to use electric microwave heaters instead of natural gas-heaters.

- **fugitive emissions**
  Unintended leaks of gas from the processing, transmission, and/or transportation of fossil fuels.

- **full forced outage**
  The net capability of main generating units that are unavailable for load for emergency reasons.

- **full power day**
  The equivalent of 24 hours of full power operation by a reactor. The number of full power days in a specific cycle is the product of the reactor's capacity factor and the length of the cycle.

- **full power operation**
  Operation of a unit at 100 percent of its design capacity. Full-power operation precedes commercial operation.

- **full requirements consumer**
  A wholesale consumer without other generating resources whose electric energy seller is the sole source of long-term firm power for the consumer's service area. The terms and conditions of sale are equivalent to the seller's obligations to its own retail service, if any.

- **fumarole**
  A vent from which gas or steam issue; a geyser or spring that emits gases.

■ **furnace coke plant**

A coke plant whose coke production is used primarily by the producing company.

■ **furnace**

The part of a boiler or warm-air space-heating plant in which combustion takes place.

■ **furnaces that heat directly, without using steam or hot water (similar to a residential furnace)**

Furnaces burn natural gas, fuel oil, propane/ butane (bottled gas), or electricity to warm the air. The warmed air is then distributed throughout the building through ducts. Many people use the words "boilers" and "furnaces" interchangeably. They are not the same. We mean that warm air is produced directly by burning some fuel. Costs of labor to operate the wells and related equipment and facilities; repair and maintenance costs; the costs of materials, supplies, and fuels consumed and services utilized in operating the wells and related equipment and facilities; the costs of property taxes and insurance applicable to proved properties and wells and related equipment and facilities; the costs of severance taxes. Depreciation, depletion, and amortization (DD&A) of capitalized acquisition, exploration, and development costs are not production costs, but also become part of the cost of oil and gas produced along with production (lifting) costs identified above. Production costs include the following subcategories of costs:

- well workers and maintenance;
- operating fluid injections and improved recovery programs;
- operating gas processing plants;
- ad valorem taxes;
- production or severance taxes;
- other, including overhead.

■ **futures market**

A trade center for quoting prices on contracts for the delivery of a specified quantity of a commodity at a specified time and place in the future.

■ **gallon**

A volumetric measure equal to 4 quarts (231 cubic inches) used to measure fuel oil. One barrel equals 42 gallons.

■ **gas**

A non-solid, non-liquid combustible energy source that includes natural gas, coke-oven gas, blast-furnace gas, and refinery gas.

■ **gas cooled fast breeder reactor (gcfb)**

A fast breeder reactor that is cooled by a gas (usually helium) under pressure.

gas oil
European and Asian designation for No. 2 heating oil and No. 2 diesel fuel.

gas plant operator
Any firm, including a gas plant owner, which operates a gas plant and keeps the gas plant records. A gas plant is a facility in which natural gas liquids are separated from natural gas or in which natural gas liquids are fractionated or otherwise separated into natural gas liquid products or both.

gas processing unit
A facility designed to recover natural gas liquids from a stream of natural gas that may or may not have passed through lease separators and/or field separation facilities. Another function of natural gas processing plants is to control the quality of the processed natural gas stream. Cycling plants are considered natural gas processing plants.

gas to liquids (gtl)
A process that combines the carbon and hydrogen elements in natural gas molecules to make synthetic liquid petroleum products, such as diesel fuel.

gas turbine plant
A plant in which the prime mover is a gas turbine. A gas turbine consists typically of an axial-flow air compressor and one or more combustion chambers where liquid or gaseous fuel is burned and the hot gases are passed to the turbine and where the hot gases expand drive the generator and are then used to run the compressor.

gas well
A well completed for production of natural gas from one or more gas zones or reservoirs. Such wells contain no completions for the production of crude oil.

gas well productivity
Derived annually by dividing gross natural gas withdrawals from gas wells by the number of producing gas wells on December 31 and then dividing the quotient by the number of days in the year.

gasification
A method for converting coal, petroleum, biomass, wastes, or other carbon-containing materials into a gas that can be burned to generate power or processed into chemicals and fuels.

gasohol
A blend of finished motor gasoline containing alcohol (generally ethanol but sometimes methanol) at a concentration between 5.7 percent and 10 percent by volume.

gasoline blending components
Naphthas which will be used for blending or compounding into

finished aviation or motor gasoline (e.g., straight-run gasoline, alkylate, reformate, benzene, toluene, and xylene). Excludes oxygenates (alcohols, ethers), butane, and pentanes plus.

▪ **gasoline grades**
The classification of gasoline by octane ratings. Each type of gasoline (conventional, oxygenated, and reformulated) is classified by three grades - Regular, Midgrade, and Premium. Note: Gasoline sales are reported by grade in accordance with their classification at the time of sale. In general, automotive octane requirements are lower at high altitudes. Therefore, in some areas of the United States, such as the Rocky Mountain States, the octane ratings for the gasoline grades may be 2 or more octane points lower.

- **Regular gasoline**:Gasoline having an antiknock index, i.e., octane rating, greater than or equal to 85 and less than 88. Note: Octane requirements may vary by altitude.
- **Midgrade gasoline**: Gasoline having an antiknock index, i.e., octane rating, greater than or equal to 88 and less than or equal to 90. Note: Octane requirements may vary by altitude.
- **Premium gasoline**: Gasoline having an antiknock index, i.e., octane rating, greater than 90. Note: Octane requirements may vary by altitude.

▪ **gasoline motor, (leaded)**
Contains more than 0.05 grams of lead per gallon or more than 0.005 grams of phosphorus per gallon. The actual lead content of any given gallon may vary. Premium and regular grades are included, depending on the octane rating. Includes leaded gasohol. Blendstock is excluded until blending has been completed. Alcohol that is to be used in the blending of gasohol is also excluded.

▪ **gasoline motor, (unleaded)**
Contains not more than 0.05 grams of lead per gallon and not more than 0.005 grams of phosphorus per gallon. Premium and regular grades are included, depending on the octane rating. Includes unleaded gasohol. Blendstock is excluded until blending has been completed. Alcohol that is to be used in the blending of gasohol is also excluded.

▪ **gate station**
Location where the pressure of natural gas being transferred from the transmission system to the distribution system is lowered for transport through small diameter, low pressure pipelines.

▪ **gatherer**
A company primarily engaged in the gathering of natural gas from

well or field lines for delivery, for a fee, to a natural gas processing plant or central point. Gathering companies may also provide compression, dehydration, and/or treating services.

- **generally accepted accounting principles (gaap)**
Defined by the FASB as the conventions, rules, and procedures necessary to define accepted accounting practice at a particular time, includes both broad guidelines and relatively detailed practices and procedures.

- **generating facility**
An existing or planned location or site at which electricity is or will be produced.

- **generating station**
A station that consists of electric generators and auxiliary equipment for converting mechanical, chemical, or nuclear energy into electric energy.

- **generating unit**
Any combination of physically connected generators, reactors, boilers, combustion turbines, and other prime movers operated together to produce electric power.

- **generation company**
An entity that owns or operates generating plants. The generation company may own the generation plants or interact with the short-term market on behalf of plant owners.

- **generation**
The process of producing electric energy by transforming other forms of energy; also, the amount of electric energy produced, expressed in kilowatt hours.

- **generator capacity**
The maximum output, commonly expressed in megawatts (MW), that generating equipment can supply to system load, adjusted for ambient conditions.

- **generator nameplate capacity (installed)**
The maximum rated output of a generator, prime mover, or other electric power production equipment under specific conditions designated by the manufacturer. Installed generator nameplate capacity is commonly expressed in megawatts (MW) and is usually indicated on a nameplate physically attached to the generator.

- **geologic assurance**
State of sureness, confidence, or certainty of the existence of a quantity of resources based on the distance from points where coal is measured or sampled and on the abundance and quality of geologic data as related to thickness of overburden, rank, quality, thickness of coal, a real

extent, geologic history, structure, and correlations of coal beds and enclosing rocks. The degree of assurance increases as the nearness to points of control, abundance, and quality of geologic data increases.

- **geologic considerations**
Conditions in the coal deposit or in the rocks in which it occurs that may complicate or preclude mining. Geologic considerations are evaluated in the context of the current state of technology and regulations, so the impact on mining may change with time.

- **geological and geophysical (g&g) costs**
Costs incurred in making geological and geophysical studies, including, but not limited to, costs incurred for salaries, equipment, obtaining rights of access, and supplies for scouts, geologists, and geophysical crews.

- **geological oil formation**

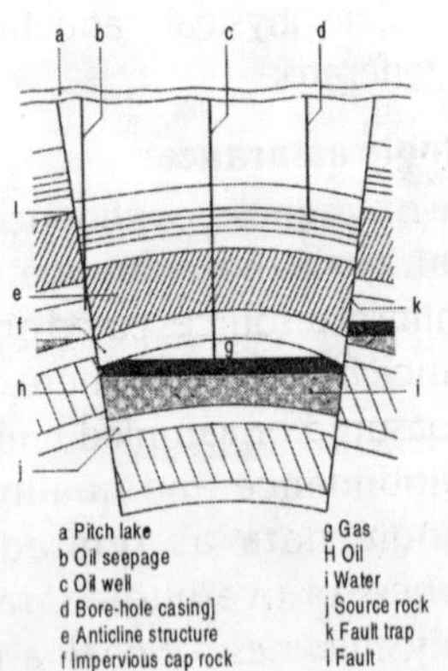

- **geological repository**
A mined facility for disposal of radioactive waste that uses waste packages and the natural geology as barriers to provide waste isolation.

- **geopressured**
A type of geothermal resource occurring in deep basins in which the fluid is under very high pressure.

- **geothermal energy**
Geothermal Energy has been around for as long as the Earth has existed. "Geo" means earth, and "thermal" means heat. So, geothermal means earth-heat.

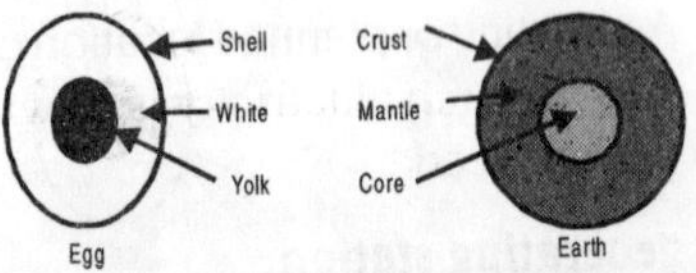

Have you ever cut a boiled egg in half? The egg is similar to how the earth looks like inside. The yellow yolk of the egg is like the core of the earth. The white part is the mantle of the earth. And the thin shell of the egg, that would have surrounded the boiled egg if you didn't peel it off, is like the earth's crust.

- **geothermal energy**
Hot water or steam extracted from geothermal reservoirs in the earth's crust. Water or steam extracted from geothermal reservoirs can be used for geothermal heat pumps, water heating, or electricity generation.

geothermal plant
A plant in which the prime mover is a steam turbine. The turbine is driven either by steam produced from hot water or by natural steam that derives its energy from heat found in rocks or fluids at various depths beneath the surface of the earth. The energy is extracted by drilling and/or pumping.

geyser
A special type of thermal spring that periodically ejects water with great force.

giga
One billion.

gigawatt (gw)
One billion watts or one thousand megawatts.

gigawatt-electric (gwe)
One billion watts of electric capacity.

gigawatthour (gwh)
One billion watthours.

gilsonite
Trademark name for uintaite (or uintahite), a black, brilliantly lustrous natural variety of asphalt found in parts of Utah and western Colorado.

global warming
An increase in the near surface temperature of the Earth. Global warming has occurred in the distant past as the result of natural influences, but the term is today most often used to refer to the warming some scientists predict will occur as a result of increased anthropogenic emissions of greenhouse gases.

global warming potential (gwp)
An index used to compare the relative radiative forcing of different gases without directly calculating the changes in atmospheric concentrations. GWPs are calculated as the ratio of the radiative forcing that would result from the emission of one kilogram of a greenhouse gas to that from the emission of one kilogram of carbon dioxide over a fixed period of time, such as 100 years.

government-owned stocks
Oil stocks owned by the national government and held for national security. In the United States, these stocks are known as the Strategic Petroleum Reserve.

green pricing
In the case of renewable electricity, green pricing represents a market solution to the various problems associated with regulatory valuation of the non-market benefits of renewables. Green pricing programs allow electricity customers to express their willingness to pay for renewable energy development through

direct payments on their monthly utility bills.

■ greenhouse effect
The result of water vapor, carbon dioxide, and other atmospheric gases trapping radiant (infrared) energy, thereby keeping the earth's surface warmer than it would otherwise be. Greenhouse gases within the lower levels of the atmosphere trap this radiation, which would otherwise escape into space, and subsequent re-radiation of some of this energy back to the Earth maintains higher surface temperatures than would occur if the gases were absent.

■ greenhouse gases
Those gases, such as water vapor, carbon dioxide, nitrous oxide, methane, hydrofluorocarbons (HFCs), perfluorocarbons (PFCs) and sulfur hexafluoride, that are transparent to solar (short-wave) radiation but opaque to long-wave (infrared) radiation, thus preventing long-wave radiant energy from leaving Earth's atmosphere. The net effect is a trapping of absorbed radiation and a tendency to warm the planet's surface.

■ grid
The layout of an electrical distribution system.

■ gross additions to construction work in progress for the month
This amount should include the monthly gross additions for an electric plant in the process of construction.

■ gross company-operated production
Total production from all company-operated properties, including all working and nonworking interests.

■ gross domestic product (gdp) implicit price deflator
The implicit price deflator, published by the U.S. Department of Commerce, Bureau of Economic Analysis, is used to convert nominal figures to real figures.

■ gross domestic product (gdp)
The total value of goods and services produced by labor and property located in the United States. As long as the labor and property are located in the United States, the supplier (that is, the workers and, for property, the owners) may be either U.S. residents or residents of foreign countries.

■ gross energy intensity
Total consumption of a particular energy source(s) or fuel(s) by a group of buildings, divided by the total floor space of those buildings,

including buildings and floor space where the energy source or fuel is not used, i.e., the ratio of consumption to gross floor space.

- **gross gas withdrawal**
  The full-volume of compounds extracted at the wellhead, including non-hydrocarbon gases and natural gas plant liquids.

- **gross generation**
  The total amount of electric energy produced by generating units and measured at the generating terminal in kilowatt hours (kWh) or megawatt hours (MWh).

- **gross head**
  A dam's maximum allowed vertical distance between the upstream's surface water (headwater) forebay elevation and the down stream's surface water (tailwater) elevation at the tail-race for reaction wheel dams or the elevation of the jet at impulse wheel dams during specified operation and water conditions.

- **gross input to atmospheric crude oil distillation units**
  Total input to atmospheric crude oil distillation units. Includes all crude oil, lease condensate, natural gas plant liquids, unfinished oils, liquefied refinery gases, slop oils, and other liquid hydrocarbons produced from tar sands, gilsonite, and oil shale.

- **gross inputs**
  The crude oil, unfinished oils, and natural gas plant liquids put into atmospheric crude oil distillation units.

- **gross national product (gnp)**
  The total value of goods and services produced by the nation's economy before deduction of depreciation charges and other allowances for capital consumption. It includes the total purchases of goods and services by private consumers and government, gross private domestic capital investment, and net foreign trade.

- **gross vehicle weight rating (gvwr)**
  Vehicle weight plus carrying capacity.

- **gross withdrawals**
  Full well stream volume, including all natural gas plant liquid and nonhydrocarbon gases, but excluding lease condensate. Also includes amounts delivered as royalty payments or consumed in field operations.

- **gross working interest ownership basis**
  Gross working interest ownership is the respondent's working interest in a given property plus the proportionate share of any royalty interest, including overriding royalty interest, associated with the working interest.

group

A group is a logical grouping of assemblies with similar characteristics. All assemblies in a group have the same initial average enrichment, the same cycle/reactor history, the same current location, the same burn up, the same owner, and the same assembly type.

group name

The DOE/EIA-assigned name identifying a composite supply source (i.e., commonly metered gas streams from more than one field), which is often the case in contract areas, field areas, and plants. A group name can also be a pipeline purchase (i.e., FERC Gas Tariff, Canadian Gas, Mexican Gas, and Algerian LNG). Emergency purchases and short term purchases are also group names. Group Code - The DOE/ EIA-assigned code identifying a composite supply source.

group quarters

Living arrangement for institutional groups containing ten or more unrelated persons. Group quarters are typically found in hospitals, nursing or rest homes, military barracks, ships, halfway houses, college dormitories, fraternity and sorority houses, convents, monasteries, shelters, jails, and correctional institutions. Group quarters may also be found in houses or apartments shared by ten or more unrelated persons. Group quarters are often equipped with a dining area for residents.

gypsum

Calcium sulfate dihydrate (C_aSO_4 $2H_2O$) a sludge constituent from the conventional lime scrubber process, obtained as a byproduct of the dewatering operation and sold for commercial use.

half-life

The time it takes for an isotope to lose half of its radioactivity.

halogen lamp

A type of incandescent lamp that lasts much longer and is more efficient than the common incandescent lamp. The lamp uses a halogen gas, usually iodine or bromine, that causes the evaporating tungsten to be redeposited on the filament, thus prolonging its life.

halogenated substances

A volatile compound containing halogens, such as chlorine, fluorine or bromine.

hand loading

An underground loading method by which coal is removed from the working face by manual labor through the use of a shovel for conveyance to the surface. Though rapidly disappearing, it is still used in small-tonnage mines.

haulage cost

Cost of loading ore at a mine site and transporting it to a processing plant.

head

The product of the water's weight and a usable difference in elevation gives a measurement of the potential energy possessed by water.

heap leach solutions

The separation, or dissolving-out from mined rock of the soluble uranium constituents by the natural action of percolating a prepared chemical solution through mounded (heaped) rock material. The mounded material usually contains low grade mineralized material and/or waste rock produced from open pit or underground mines. The solutions are collected after percolation is completed and processed to recover the valued components.

heat content

Measurement: The gross heat content (or heating value), is the number of British thermal units (Btu) produced by the combustion, at constant pressure, of the amount of the gas that would occupy a volume of one cubic foot at a temperature of 60 degrees Fahrenheit, if saturated with water vapor and under a pressure equivalent to 30 inches of mercury at 32 degrees Fahrenheit and under standard gravitational force (980.665 cm per sec.^{2}), with air of the same temperature and pressure as the gas, when the products of combustion are cooled to the initial temperature of gas and air and when the water formed by combustion is condensed to the liquid state.

heat pump (air source)

An air-source heat pump is the most common type of heat pump. The heat pump absorbs heat from the outside air and transfers the heat to the space to be heated in the heating mode. In the cooling mode the heat pump absorbs heat from the space to be cooled and rejects the heat to the outside air. In the heating mode when the outside air approaches 32o F or less, air-source heat pumps loose efficiency and generally require a back-up (resistance) heating system.

heat pump (geothermal)

A heat pump in which the refrigerant exchanges heat (in a heat exchanger) with a fluid circulating through an earth connection medium (ground or ground water). The fluid is contained in a variety of loop (pipe) configurations depending on the temperature of the ground and the ground area available. Loops may be installed horizontally or vertically in the ground or submersed in a body of water.

heat pump efficiency

The efficiency of a heat pump, that is, the electrical energy to operate it, is directly related to temperatures between which it operates. Geothermal heat pumps are more efficient than conventional heat pumps or air conditioners that use the outdoor air since the ground or ground water a few feet below the earth's surface remains relatively constant throughout the year. It is more efficient in the winter to draw heat from the relatively warm ground than from the atmosphere where the air temperature is much colder, and in summer transfer waste heat to the relatively cool ground than to hotter air. Geothermal heat pumps are generally more expensive ($2,000-$5,000) to install than outside air heat pumps. However, depending on the location geothermal heat pumps can reduce energy consumption (operating cost) and correspondingly, emissions by more than 20 percent compared to high-efficiency outside air heat pumps. Geothermal heat pumps also use the waste heat from air-conditioning to provide free hot water heating in the summer.

- **heat pump**

Heating and/or cooling equipment that, during the heating season, draws heat into a building from outside and, during the cooling season, ejects heat from the building to the outside. Heat pumps are vapor-compression refrigeration systems whose indoor/outdoor coils are used reversibly as condensers or evaporators, depending on the need for heating or cooling.

- **heat rate**

A measure of generating station thermal efficiency, generally expressed in Btu per net kilowatt hour. It is computed by dividing the total Btu content of fuel burned for electric generation by the resulting net kilowatt hour generation.

- **heated floor space**

The area within a building that is space heated.

- **heating degree-days (hdd)**

A measure of how cold a location is over a period of time relative to a base temperature, most commonly specified as 65 degrees Fahrenheit. The measure is computed for each day by subtracting the average of the day's high and low temperatures from the base temperature (65 degrees), with negative values set equal to zero. Each day's heating degree-days are summed to create a heating degree-day measure for a specified reference period. Heating degree-days are used in energy analysis as an indicator of space heating energy requirements or use.

- **heating equipment**

Any equipment designed and/or specifically used for heating

ambient air in an enclosed space. Common types of heating equipment include: central warm air furnace, heat pump, plug-in or built-in room heater, boiler for steam or hot water heating system, heating stove, and fireplace.

- **heating intensity**

The ratio of space-heating consumption or expenditures to square footage of heated floor space and heating degree-days (base 65 degrees Fahrenheit). This ratio provides a way of comparing different types of housing units and households by controlling for differences in housing unit size and weather conditions. The square footage of heated floor space is based on the measurements of the floor space that is heated. The ratio is calculated on a weighted, aggregate basis according to the following formula: Heating Intensity = Btu for Space Heating / (Heated Square Feet * Heating Degree-Days).

- **heating stove burning wood, coal, or coke**

Any free-standing box or controlled-draft stove; or a stove installed in a fireplace opening, using the chimney of the fireplace. Stoves are made of cast iron, sheet metal, or plate steel. Free-standing fireplaces that can be detached from their chimneys are considered heating stoves.

- **heating value**

The average number of British thermal units per cubic foot of natural gas as determined from tests of fuel samples.

- **heavy gas oil**

Petroleum distillates with an approximate boiling range from 651 degrees to 1000 degrees Fahrenheit.

- **heavy metals**

Metallic elements, including those required for plant and animal nutrition, in trace concentration but which become toxic at higher concentrations. Examples are mercury, chromium, cadmium, and lead.

- **heavy oil**

The fuel oils remaining after the lighter oils have been distilled off during the refining process. Except for start-up and flame stabilization, virtually all petroleum used in steam plants is heavy oil. Includes fuel oil numbers 4, 5, and 6; crude; and topped crude.

- **heavy rail**

An electric railway with the capacity for a "heavy volume" of traffic and characterized by exclusive rights-of-way, multi-car trains, high speed and rapid acceleration, sophisticated signaling, and high platform loading. Also known as "subway," elevated (railway), "metropolitan railway (metro)."

- **heavy water**

Water containing a significantly greater proportion of heavy hydrogen (deuterium) atoms to ordinary hydrogen atoms than is found in ordinary (light) water. Heavy water is used as a moderator in some reactors because it slows neutrons effectively and also has a low cross section for absorption of neutrons.

- **heavy-water-moderated reactor**

A reactor that uses heavy water as its moderator. Heavy water is an excellent moderator and thus permits the use of inexpensive natural (unenriched) uranium as fuel.

- **hedging contracts**

Contracts which establish future prices and quantities of electricity independent of the short-term market.

- **hedging**

The buying and selling of futures contracts so as to protect energy traders from unexpected or adverse price fluctuations.

- **heliostat**

A mirror that reflects solar rays onto a central receiver. A heliostat automatically adjusts its position to track daily or seasonal changes in the sun's position. The arrangement of heliostats around a central receiver is also called a solar collector field.

- **henry hub**

A pipeline hub on the Louisiana Gulf coast. It is the delivery point for the natural gas futures contract on the New York Mercantile Exchange (NYMEX).

- **high efficiency ballast**

A lighting conservation feature consisting of an energy-efficient version of a conventional electromagnetic ballast. The ballast is the transformer for fluorescent and high-intensity discharge (HID) lamps, which provides the necessary current, voltage, and wave-form conditions to operate the lamp. A high-efficiency ballast requires lower power input than a conventional ballast to operate HID and fluorescent lamps.

- **high efficiency lighting**

Lighting provided by high-intensity discharge (HID) lamps and/or fluorescent lamps.

- **high-intensity discharge (hid) lamp**

A lamp that produces light by passing electricity through gas, which causes the gas to glow. Examples of HID lamps are mercury vapor lamps, metal halide lamps, and high-pressure sodium lamps. HID lamps have extremely long life and emit far more lumens per fixture than do fluorescent lights.

- **high-mileage households**
  Households with estimated aggregate annual vehicle mileage that exceeds 12,500 miles.

- **high-temperature collector**
  A solar thermal collector designed to operate at a temperature of 180 degrees Fahrenheit or higher.

- **highwall**
  The unexcavated face of exposed over-burden and coal in a surface mine.

- **hinshaw pipeline**
  A pipeline or local distribution company that has received exemptions from regulations pursuant to the Natural Gas Act. These companies transport interstate natural gas not subject to regulations under NGA.

- **holding company**
  A company that confines its activities to owning stock in and supervising management of other companies. The Securities and Exchange Commission, as administrator of the Public Utility Holding Company Act of 1935, defines a holding company as "a company which directly or indirectly owns, controls or holds 10 percent or more of the outstanding voting securities of a holding company".

- **holding pond**
  A structure built to contain large volumes of liquid waste to ensure that it meets environmental requirements prior to release.

- **horizontal axis wind turbine**
  The most common type of wind turbine where the axis of rotation is oriented horizontally.

- **horsepower**
  A unit for measuring the rate of work (or power) equivalent to 33,000 foot-pounds per minute or 746 watts.

- **host government**
  The government (including any government-controlled firm engaged in the production, refining, or marketing of crude oil or petroleum products) of the foreign country in which the crude oil is produced.

- **hot dry rock**
  Heat energy residing in impermeable, crystalline rock. Hydraulic fracturing may be used to create permeability to enable circulation of water and removal of the heat.

- **hot tub**
  Water-filled wood, plastic, or ceramic container in which up to 12 people can lounge. Normally equipped with a heater that heats the water from 80 degrees to 106 degrees Fahrenheit. It may also have jets to bubble the water. The

water is not drained after each use. An average-size hot tub holds 200 to 400 gallons of water. All reported hot tubs are assumed to include an electric pump. These are also called spas or jacuzzis.

- **hours under load**

The hours the boiler is operating to drive the generator producing electricity.

- **household**

A family, an individual, or a group of up to nine unrelated persons occupying the same housing unit. "Occupy" means that the housing unit is the person's usual or permanent place of residence.

- **household energy expenditures**

The total amount of funds spent for energy consumed in, or delivered to, a housing unit during a given period of time.

- **housing unit**

A house, an apartment, a group of rooms, or a single room if it is either occupied or intended for occupancy as separate living quarters by a family, an individual, or a group of one to nine unrelated persons. Separate living quarters means the occupants (1) live and eat separately from other persons in the house or apartment and (2) have direct access from the outside of the buildings or through a common hall—that is, they can get to it without going through someone else's living quarters. Housing units do not include group quarters such as prisons or nursing homes where ten or more unrelated persons live. A common dining area used by residents is an indication of group quarters. Hotel and motel rooms are considered housing units if occupied as the usual or permanent place of residence.

- **hub height**

In a horizontal-axis wind turbine, the distance from the turbine platform to the rotor shaft.

- **humidifier**

A humidifier adds moisture to the air (often needed in winter when indoor air is very dry). It may be a portable unit or attached to the heating system.

- **humidity**

The moisture content of air. Relative humidity is the ratio of the amount of water vapor actually present in the air to the greatest amount possible at the same temperature.

- **hvac**

An abbreviation for the heating, ventilation, and air-conditioning system; the system or systems that condition air in a building.

- **hvac conservation feature**

A building feature designed to reduce the amount of energy consumed by the heating, cooling, and ventilating equipment.

- **hvac dsm program**

A DSM (demand-side management) program designed to promote the efficiency of the heating or cooling delivery system, including replacement. Includes ventilation (economizers; heat recovery from exhaust air), cooling (evaporative cooling, cool storage; heat recovery from chillers; high-efficiency air conditioning), heating, and automatic energy management systems.

- **hybrid transmission line**

A double-circuit line that has one alternating current and one direct circuit. The AC circuit usually serves local loads along the line.

- **hydraulic fracturing**

Fracturing of rock at depth with fluid pressure. Hydraulic fracturing at depth may be accomplished by pumping water into a well at very high pressures. Under natural conditions, vapor pressure may rise high enough to cause fracturing in a process known as hydrothermal brecciation.

- **hydraulic head**

The distance between the respective elevations of the upstream water surface (headwater) above and the downstream surface water (tail water) below a hydroelectric power plant.

- **hydro power**

When it rains in hills and mountains, the water becomes streams and rivers that run down to the ocean. The moving or falling water can be used to do work. Energy, you'll remember is the ability to do work. So moving water, which has kinetic energy, can be used to make electricity. For hundreds of years, moving water was used to turn wooden wheels that were attached to grinding wheels to grind (or mill) flour or corn. These were called grist mills or water mills.

- **hydrocarbon**

An organic chemical compound of hydrogen and carbon in the gaseous, liquid, or solid phase. The molecular structure of hydrocarbon compounds varies from the simplest (methane, a constituent of natural gas) to the very heavy and very complex.

- **hydrochlorofluorocarbons (hcfcs)**

Chemicals composed of one or more carbon atoms and varying numbers of hydrogen, chlorine, and fluorine atoms.

- **hydroelectric power**

Water is needed to run a hydroelectric generating unit. It's held in a reservoir or lake behind the dam, and the force of the water being released from the reservoir through the dam spins the blades of a turbine. The turbine is connected to the generator that produces electricity. After passing through the turbine, the water reenters the river on the downstream side of the dam.

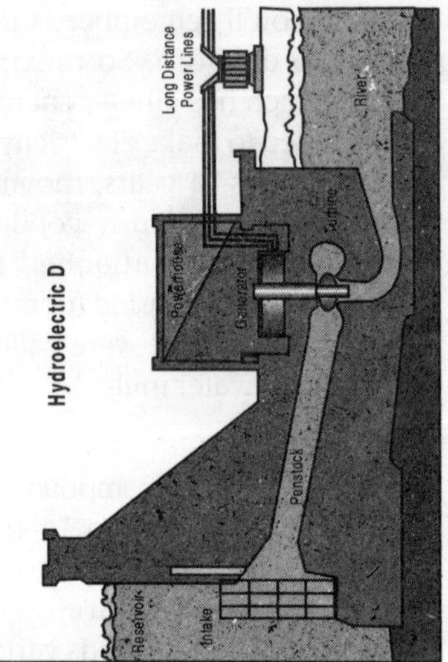

Theuse of flowing water to produce electrical energy.

- **hydrofluorocarbons (hfcs)**

A group of man-made chemicals composed of one or two carbon atoms and varying numbers of hydrogen and fluorine atoms. Most HFCs have 100-year Global Warming Potentials in the thousands.

- **hydrogen**

A colorless, odorless, highly flammable gaseous element. It is the lightest of all gases and the most abundant element in the universe, occurring chiefly in combination with oxygen in water and also in acids, bases, alcohols, petroleum, and other hydrocarbons.

- **hydroxyl radical (oh)**

An important chemical scavenger of many trace gases in the atmosphere that are greenhouse gases. Atmospheric concentrations of OH affect the atmospheric lifetimes of greenhouse gases, their abundance, and, ultimately, the effect they have on climate.

- **hypothetical resources (coal)**

Undiscovered coal resources in beds that may reasonably be expected to exist in known mining districts under known geologic conditions. In general, hypothetical resources are in broad areas of coalfields where points of observation are absent and evidence is from distant outcrops, drill holes, or wells. Exploration that confirms their existence and better defines their quantity and quality would permit their reclassification as identified resources. Quantitative estimates are based on a broad knowledge of the geologic character of coalbed or region. Measurements of coal thickness are more than 6 miles apart. The assumption of continuity of coalbed is supported only by geologic evidence.

- **identified resources**

Coal deposits whose location, rank, quality, and quantity are known from geologic evidence supported by engineering measurements. Included are beds of bituminous coal and anthracite (14 or more inches thick) and beds of subbituminous coal and lignite (30 or more inches thick) that occur at depths to 6,000 feet. The existence and quantity of these beds have been delineated within specified degrees of geologic assurance as measured, indicated, or inferred. Also included are thinner and/or deeper beds that presently are being mined or for which there is evidence that they could be mined commercially.

- **idle capacity**

The component of operable capacity that is not in operation and not under active repair, but capable of being placed in operation within 30 days; and capacity not in operation but under active repair that can be completed within 90 days.

- **international energy agency**

If a state agency uses a different basis for classifying onshore and offshore areas, the state classification should be used (e.g., Cook Inlet in Alaska is classified as offshore; for Louisiana, the coastline is defined as the Chapman Line, as modified by subsequent adjudication).

- **impedance**

The opposition to power flow in an AC circuit. Also, any device that introduces such opposition in the form of resistance, reactance, or both. The impedance of a circuit or device is measured as the ratio of voltage to current, where a sinusoidal voltage and current of the same frequency are used for the measurement; it is measured in ohms.

- **implicit price deflator**

The implicit price deflator, published by the U.S. Department of Commerce, Bureau of Economic Analysis, is used to convert nominal figures to real figures.

- **imported crude oil burned as fuel**

The amount of foreign crude oil burned as a fuel oil, usually as residual fuel oil, without being processed as such. Imported crude oil burned as fuel includes lease condensate and liquid hydrocarbons produced from tar sands, gilsonite, and oil shale.

- **imported refiners' acquisition cost (irac)**

The average price for imported oil paid by U.S. refiners.

- **imports**

Receipts of goods into the 50 States and the District of Columbia from U.S. possessions and territories or from foreign countries.

- **improved recovery**

Extraction of crude oil or natural gas by any method other than those that rely primarily on natural reservoir pressure, gas lift, or a system of pumps.

- **in situ leach mining (isl)**

The recovery, by chemical leaching, of the valuable components of a mineral deposit without physical extraction of the mineralized rock from the ground. Also referred to as "solution mining."

- **inadvertent power exchange**

An unintended power exchange among utilities that is either not previously agreed upon or in an amount different from the amount agreed upon.

- **incandescent lamp**

A glass enclosure in which light is produced when a tungsten filament is electrically heated so that it glows. Much of the energy is converted into heat; therefore, this class of lamp is a relatively inefficient source of light. Included in this category are the familiar screw-in light bulbs, as well as somewhat more efficient lamps, such as tungsten halogen lamps, reflector or r-lamps, parabolic aluminized reflector (PAR) lamps, and ellipsoidal reflector (ER) lamps.

- **incandescent light bulbs, including regular or energy-efficient light bulbs**

An incandescent bulb is a type of electric light in which light is produced by a filament heated by electric current. The most common example is the type you find in most table and floor lamps. In commercial buildings, incandescent lights are used for display lights in retail stores, hotels and motels. This includes the very small, high-intensity track lights used to display merchandise or provide spot illumination in restaurants. Energy efficient light bulbs, known as "watt-savers," use less energy than a standard incandescent bulb. "Long-life" bulbs, bulbs that last longer than standard incandescent but produce considerably less light, are not considered energy-efficient bulbs. This category also includes halogen lamps. Halogen lamps are a special type of incandescent lamp containing halogen gas to produce a brighter, whiter light than standard incandescent. Halogen lamps come in three styles: bulbs, models with reflectors, and infrared models with reflectors. Halogen lamps are especially suited to recessed or "canned fixtures," track lights, and outdoor lights.

- **incentives demand-side management (dsm) program assistance**

This DSM program assistance offers monetary or non-monetary awards to encourage consumers to buy energy-efficient equipment and to participate in programs designed to reduce energy usage. Examples of incentives are zero or low-interest loans, rebates, and direct installation of low cost measures, such as water heater wraps or duct work for distributing the cool air; the units condition air only in the room or areas where they are located.

- **incremental effects**

The annual changes in energy use (measured in megawatt hours) and peak load (measured in kilowatts) caused by new participants in existing DSM (Demand-Side Management) programs and all participants in new DSM programs during a given year. Reported Incremental Effects are annualized to indicate the program effects that would have occurred had these participants been initiated into the program on January 1 of the given year. Incremental effects are not simply the Annual Effects of a given year minus the Annual Effects of the prior year, since these net effects would fail to account for program attrition, equipment degradation, building demolition, and participant dropouts. Please note that Incremental Effects are not a monthly disaggregate of the Annual Effects, but are the total year's effects of only the new participants and programs for that year.

- **incremental energy costs**

The additional cost of producing and/or transmitting electric energy above some previously determined base cost.

- **independent power producer**

A corporation, person, agency, authority, or other legal entity or instrumentality that owns or operates facilities for the generation of electricity for use primarily by the public, and that is not an **electric utility**.

- **independent system operator (iso)**

An independent, Federally regulated entity established to coordinate regional transmission in a non-discriminatory manner and ensure the safety and reliability of the electric system.

- **indian coal lease**

A lease granted to a mining company to produce coal from Indian lands in exchange for royalties and other revenues; obtained by direct negotiation with Indian tribal authorities, but subject to approval and administration by the U.S. Department of the Interior.

- **indicated resources, coal**

Coal for which estimates of the rank, quality, and quantity are based partly on sample analyses and measurements and partly on reasonable geologic projections. Indicated resources are computed partly from specified measurements and partly from projection of visible data for a reasonable distance on the basis of geologic evidence. The points of observation are 1/2 to 1-1/2 miles apart. Indicated coal is projected to extend as a 1/2-mile-wide belt that lies more than 1/4 mile from the outcrop, points of observation, or measurement.

▪ **indirect cost**

Costs not directly related to mining or milling operations, such as overhead, insurance, security, office expenses, property taxes, and similar administrative expenses.

▪ **indirect uses (end-use category)**

The end-use category that handles boiler fuel. Fuel in boilers is transformed into another useful energy source, steam or hot water, which is in turn used in other end uses, such as process or space heating or electricity generation. Manufacturers find measuring quantities of steam as it passes through to various end uses especially difficult because variations in both temperature and pressure affect energy content. Thus, the MECS does not present end-use estimates of steam or hot water and shows only the amount of the fuel used in the boiler to produce those secondary energy sources.

▪ **indirect utility cost**

A utility cost that may not be meaningfully identified with any particular DSM program category. Indirect costs could be attributable to one of several accounting cost categories (i.e., Administrative, Marketing, Monitoring & Evaluation, Utility-Earned Incentives, Other). Accounting costs that are known DSM program costs should not be reported under Indirect Utility Cost; those costs should be reported as Direct Utility Costs under the appropriate DSM program category.

▪ **industrial production**

The Federal Reserve Board calculates this index by compiling indices of physical output from a variety of agencies and trade groups, weighting each index by the Census' value added, and adding it to the cost of materials. When physical measures are not available, the Federal Reserve Board uses the number of production workers or amount of electricity consumed as the basis for the index. To convert industrial production into dollars, multiply by the "real value added" estimate used by the Federal Reserve Board.

▪ **industrial restrictions (coal)**

Land-use restrictions that constrain, postpone, or prohibit mining in order to meet other industrial needs or goals; for example, resources not mined due to safety concerns or due to industrial or societal priorities, such as to preserve oil or gas wells that penetrate the coal reserves; to protect surface features such as pipelines, power lines, or company facilities; or to preserve public or private assets, such as highways, railroads, parks, or buildings.

- **industrial sector**

An energy-consuming sector that consists of all facilities and equipment used for producing, processing, or assembling goods. The industrial sector encompasses the following types of activity: manufacturing (NAICS codes 31-33); agriculture, forestry, fishing and hunting (NAICS code 11); mining, including oil and gas extraction (NAICS code 21); and construction (NAICS code 23). Overall energy use in this sector is largely for process heat and cooling and powering machinery, with lesser amounts used for facility heating, air conditioning, and lighting. Fossil fuels are also used as raw material inputs to manufactured products.

- **inferred resources**

Coal in unexplored extensions of demonstrated resources for which estimates of the quality and size are based on geologic evidence and projection. Quantitative estimates are based largely on broad knowledge of the geologic character of the bed or region and where few measurements of bed thickness are available. The estimates are based primarily on an assumed continuation from demonstrated coal for which there is geologic evidence. The points of observation are 1-1/2 to 6 miles apart. Inferred coal is projected to extend as a 2-1/4-mile wide belt that lies more than 3/4 mile from the outcrop, points of observation, or measurement.

- **in-house demand-side management (dsm) program sponsor**

The building's owner or management encourages consumers in the building to improve energy efficiency, reduce energy costs, change timing or energy usage, or promote the use of a different energy source by sponsoring its own DSM programs.

- **initial enrichment**

Average enrichment for a fresh fuel assembly as specified and ordered in fuel cycle planning. This average should include axial blankets and axially and radially zoned enrichments.

- **initial operation**

First availability of a newly constructed unit to provide power to the grid. For a nuclear unit, this time is when the Full Power Operating License for the unit is received.

- **injections**
  Natural gas injected into storage reservoirs.

- **inoperable capacity**
  Generating capacity that is totally or partially out of service at the time of system peak load, either for scheduled outages (see GADS definition of "scheduled outages." These include both maintenance outages and planned outages.) or for reasons such as:
  - environmental restrictions;
  - extensive modifications or repair;

  or capacity specified as being in a mothballed state. This does not include derated portions of generating capacity.

- **instantaneous peak demand**
  The maximum demand at the instant of greatest load.

- **instantaneous water heater**
  Also called a "tankless" or "point-of-use" water heater. The water is heated at the point of use as it is needed.

- **institutional living quarters**
  Space provided by a business or organization for long-term housing of individuals whose reason for shared residence is their association with the business or organization. Such quarters commonly have both individual and group living spaces, and the business or organization is responsible for some aspects of resident life beyond the simple provision of living quarters. Examples include prisons; nursing homes and other long-term medical care facilities; military barracks; college dormitories; and convents and monasteries.

- **insulation**
  Any material or substance that provides a high resistance to the flow of heat from one surface to another. The different types include blanket or batt, foam, or loose fill, which are used to reduce heat transfer by conduction. Dead air space is an insulating medium in storm windows and storms as it reduces passage of heat through conduction and convection. Reflective materials are used to reduce heat transfer by radiation.

- **insulation around heating and/or cooling ducts**
  Extra insulation around the heating and/or cooling ducts intended to reduce the loss of hot or cold air as it travels to different parts of the residence.

- **insulation around hot-water pipes**
  Wrapping of insulating material around hot-water pipes to reduce the loss of heat through the pipes.

insulation around water heater

Blanket insulation wrapped around the water heater to reduce loss of heat. To qualify under this definition, this wrapping must be in addition to any insulation provided by the manufacturer.

insulator

A material that is a very poor conductor of electricity. The insulating material is usually a ceramic or fiberglass when used in the transmission line and is designed to support a conductor physically and to separate it electrically from other conductors and supporting material.

intangible drilling and development costs (idc)

Costs incurred in preparing well locations, drilling and deepening wells, and preparing wells for initial production up through the point of installing control valves. None of these functions, because of their nature, have salvage value. Such costs would include labor, transportation, consumable supplies, drilling tool rentals, site clearance, and similar costs.

integral collector storage (ics)

A solar thermal collector in which incident solar radiation is absorbed directly by the storage medium.

integrated demand

The summation of the continuously varying instantaneous demand averaged over a specified interval of time. The information is usually determined by examining a demand meter.

integrated gasification-combined cycle technology

Coal, water, and oxygen are fed to gasifier, which produces syngas. This medium-Btu gas is cleaned (particulates and sulfur compounds removed) and is fed to a gas turbine. The hot exhaust of the gas turbine and heat recovered from the gasification process are routed through a heat-recovery routed through a heat-recovery generator to produce steam, which drives a steam turbine to produce electricity.

intensity per hour

Total consumption of a particular fuel(s) divided by the total floor space of buildings that use the fuel(s) divided by total annual hours of operation.

intensity

The amount of a quantity per unit floor space. This method adjusts either the amount of energy consumed or expenditures spent, for the effects of various building characteristics, such as size of the building, number of workers, or number of operating hours, to facilitate comparisons of energy across time, fuels, and buildings.

- **interchange energy**
Kilowatt hours delivered to or received by one electric utility or pooling system from another. Settlement may be payment, returned in kind at a later time, or accumulated as energy balances until the end of the stated period.

- **intercity bus**
A bus designed for high speed, long distance travel; equipped with front doors only, high backed seats, and usually restroom facilities.

- **interconnected system**
A system consisting of two or more individual power systems normally operating with connecting tie lines.

- **interconnection**
Two or more electric systems having a common transmission line that permits a flow of energy between them. The physical connection of the electric power transmission facilities allows for the sale or exchange of energy.

- **interdepartmental sales**
Includes amounts charged by the electric department at tariff or other specified rates for electricity supplied by it to other utility departments.

- **interdepartmental service (electric)**
Interdepartmental service includes amounts charged by the electric department at tariff or other specified rates for electricity supplied by it to other utility departments.

- **interest coverage ratio**
The number of times that fixed interest charges were earned. It indicates the margin of safety of interest on fixed debt. The times-interest-earned ratio is calculated using net income before and after income taxes; and the credits of interest charged to construction being treated as other income. The interest charges include interest on long-term debt, interest on debt of associated companies, and other interest expenses.

- **intergovernmental panel on climate change (ipcc)**
A panel established jointly in 1988 by the World Meteorological Organization and the United Nations Environment Program to assess the scientific information relating to climate change and to formulate realistic response strategies.

- **interlocking directorates**
The holding of a significant position in management or a position on the corporate board of a utility while simultaneously holding a comparable position with another utility, or with a firm doing business with the utility.

- **intermediate grade gasoline**

A grade of unleaded gasoline with an octane rating intermediate between "regular" and "premium." Octane boosters are added to gasolines to control engine pre-ignition or "knocking" by slowing combustion rates.

- **intermediate load (electric system)**

The range from base load to a point between base load and peak. This point may be the midpoint, a percent of the peak load, or the load over a specified time period.

- **intermittent electric generator or intermittent resource**

An electric generating plant with output controlled by the natural variability of the energy resource rather than dispatched based on system requirements. Intermittent output usually results from the direct, non-stored conversion of naturally occurring energy fluxes such as solar energy, wind energy, or the energy of free-flowing rivers (that is, run-of-river hydroelectricity).

- **internal collector storage**

A solar thermal collector in which incident solar radiation is absorbed by the storage medium.

- **internal combustion plant**

A plant in which the prime mover is an internal combustion engine. An internal combustion engine has one or more cylinders in which the process of combustion takes place, converting energy released from the rapid burning of a fuel-air mixture into mechanical energy. Diesel or gas-fired engines are the principal types used in electric plants. The plant is usually operated during periods of high demand for electricity.

- **interruptible gas**

Gas sold to customers with a provision that permits curtailment or cessation of service at the discretion of the distributing company under certain circumstances, as specified in the service contract.

- **interruptible load**

This Demand-Side Management category represents the consumer load that, in accordance with contractual arrangements, can be interrupted at the time of annual peak load by the action of the consumer at the direct request of the system operator. This type of control usually involves large-volume commercial and industrial consumers. Interruptible Load does not include Direct Load Control.

- **interruptible or curtailable rate**

A special electricity or natural gas arrangement under which, in return for lower rates, the customer must either reduce energy demand on short notice or allow the electric or natural gas utility to temporarily cut off the energy supply for the utility to maintain service for higher priority users. This interruption or reduction in demand typically occurs during periods of high demand for the energy (summer for electricity and winter for natural gas).

- **interruptible power**
  Power and usually the associated energy made available by one utility to another. This transaction is subject to curtailment or cessation of delivery by the supplier in accordance with a prior agreement with the other party or under specified conditions.

- **interstate companies**
  Natural gas pipeline companies subject to Federal Energy Regulatory Commission (FERC) jurisdiction.

- **interstate pipeline**
  Any person engaged in natural gas transportation subject to the jurisdiction of Federal Energy Regulatory Commission (FERC) under the Natural Gas Act.

- **interstate pipeline purchase**
  Any gas supply contracted from and volumes purchased from other interstate pipelines, overland natural gas import purchases, and LNG, SNG, or coal gas purchases from domestic or foreign sources. Purchases from intrastate pipelines to section 311 (b) of the NGPA of 1978 and from independent producers are not included with interstate pipelines purchase.

- **intransit deliveries**
  Redeliveries to a foreign country of foreign gas received for transportation across U.S. territory, and deliveries of U.S. gas to a foreign country for transportation across its territory and redelivery to the United States.

- **intransit receipts**
  Receipts of foreign gas for transportation across U.S. territory and redelivery to a foreign country, and redeliveries to the United States of U.S. gas transported across foreign territory.

- **intrastate companies**
  Companies not subject to Federal Energy Regulatory Commission (FERC) jurisdiction.

- **intrastate pipeline**
  Any person engaged in natural gas transportation (not including gathering) that is not subject to the jurisdiction of the Commission under the Natural Gas Act (other than any such pipeline that is not subject to the jurisdiction of the Commission soley by reason of Section 1(c) of the Natural Gas Act).

- **in-use (vehicles)**
  Implies that a vehicle is:

1. Registered with the Government of one or more States, the District of Columbia, the Commonwealth of Puerto Rico, or the Virgin Islands; or
2. The vehicle is owned or operated by a Government or military organization within the United States that is not required to register vehicles with the Government agencies listed under 1 above. For example, civilian Federal vehicles are generally not required to register with the State Government in which they are assigned.

- **investment of municipality**

The investment of the municipality in its utility department, when such investment is not subject to cash settlement on demand or at a fixed future time. Include the cost of debt-free utility plant constructed or acquired by the municipality and made available for the use of the utility department, cash transferred to the utility department for working capital, and other expenditures of an investment nature.

- **investments and advances to unconsolidated affiliates**

The balance sheet account representing the cost of investments and advances to unconsolidated affiliates. Generally, affiliates that are less than 50-percent owned by a company may not be consolidated into the company's financial statements.

- **investor-owned utility (iou)**

A privately-owned electric utility whose stock is publicly traded. It is rate regulated and authorized to achieve an allowed rate of return.

- **ion exchange**

Reversible exchange of ions adsorbed on a mineral or synthetic polymer surface with ions in solution in contact with the surface. A chemical process used for recovery of uranium from solution by the interchange of ions between a solution and a solid, commonly a resin.

- **iron and steel industry**

Steel Works, Blast Furnaces (Including Coke Ovens), and Rolling Mills: Establishments primarily engaged in manufacturing hot metal, pig iron, and silvery pig iron from iron ore and iron and steel scrap; converting pig iron, scrap iron, and scrap steel into steel; and in hot-rolling iron and steel into basic shapes, such as plates, sheets, strips, rods, bars, and tubing.

- **irradiated nuclear fuel**

Nuclear fuel that has been exposed to radiation in the reactor core at any power level.

- **isobutane (c4h10)**

A normally gaseous branch-chain hydrocarbon. It is a colorless paraffinic gas that boils at a temperature of 10.9 degrees Fahrenheit. It is extracted from natural gas or refinery gas streams.

- **isobutylene (c4h8)**

An olefinic hydrocarbon recovered from refinery processes or petrochemical processes.

- **isohexane (c6h14)**

A saturated branch-chain hydrocarbon. It is a colorless liquid that boils at a temperature of 156.2 degrees Fahrenheit.

- **isomerization**

A refining process that alters the fundamental arrangement of atoms in the molecule without adding or removing anything from the original material. Used to convert normal butane into isobutane (C_4), an alkylation process feedstock, and normal pentane and hexane into isopentane (C_5) and isohexane (C_6), high-octane gasoline components.

- **isopach**

A line on a map drawn through points of equal thickness of a designated unit (such as a coal bed).

- **isopentane**

A saturated branched-chain hydrocarbon (C_5H_{12}) obtained by fractionation of natural gasoline or isomerization of normal pentane.

- **isotopes**

Forms of the same chemical element that differ only by the number of neutrons in their nucleus. Most elements have more than one naturally occurring isotope. Many isotopes have been produced in reactors and scientific laboratories.

- **jacket**

The enclosure on a water heater, furnace, or boiler.

- **jet fuel**

A refined petroleum product used in jet aircraft engines. It includes kerosene-type jet fuel and naphtha-type jet fuel.

- **joint implementation (ji)**

Agreements made between two or more nations under the auspices of the Framework Convention on Climate Change (FCCC) whereby a developed country can receive "emissions reduction units" when it helps to finance projects that reduce net emissions in another developed country (including countries with economies in transition).

- **joint-use facility**

A multiple-purpose hydroelectric plant. An example is a dam that stores water for both flood control and power production.

▪ joule (j)

The meter-kilogram-second unit of work or energy, equal to the work done by a force of one newton when its point of application moves through a distance of one meter in the direction of the force; equivalent to 107 ergs and one watt-second.

▪ joule's law

The rate of heat production by a steady current in any part of an electrical circuit that is proportional to the resistance and to the square of the current, or, the internal energy of an ideal gas depends only on its temperature.

▪ junction

A region of transition between semiconductor layers, such as a p/n junction, which goes from a region that has a high concentration of acceptors (p-type) to one that has a high concentration of donors (n-type).

▪ jurisdictional utilities

Utilities regulated by public laws.

▪ kaplan turbine

A type of turbine that that has two blades whose pitch is adjustable. The turbine may have gates to control the angle of the fluid flow into the blades.

▪ kerosene

A light petroleum distillate that is used in space heaters, cook stoves, and water heaters and is suitable for use as a light source when burned in wick-fed lamps. Kerosene has a maximum distillation temperature of 400 degrees Fahrenheit at the 10-percent recovery point, a final boiling point of 572 degrees Fahrenheit, and a minimum flash point of 100 degrees Fahrenheit. Included are No. 1-K and No. 2-K, the two grades recognized by ASTM Specification D 3699 as well as all other grades of kerosene called range or stove oil, which have properties similar to those of No. 1 fuel oil.

▪ kerosene-type jet fuel

A kerosene-based product having a maximum distillation temperature of 400 degrees Fahrenheit at the 10-percent recovery point and a final maximum boiling point of 572 degrees Fahrenheit and meeting ASTM Specification D 1655 and Military Specifications MIL-T-5624P and MIL-T-83133D (Grades JP-5 and JP-8). It is used for commercial and military turbojet and turboprop aircraft engines.

▪ ketone-alcohol (cyclohexanol)

An oily, colorless, hygroscopic liquid with a camphor-like odor. Used in soap making, dry cleaning, plasticizers, insecticides, and germicides.

kilovolt-ampere (kva)

A unit of apparent power, equal to 1,000 volt-amperes; the mathematical product of the volts and amperes in an electrical circuit.

kilowatt (kw)

One thousand watts.

kilowatt-electric (kwe)

One thousand watts of electric capacity.

kilowatthour (kwh)

A measure of electricity defined as a unit of work or energy, measured as 1 kilowatt (1,000 watts) of power expended for 1 hour. One kWh is equivalent to 3,412 Btu.

kinetic energy

Energy available as a result of motion that varies directly in proportion to an object's mass and the square of its velocity.

kyoto protocol

The result of negotiations at the third Conference of the Parties (COP-3) in Kyoto, Japan, in December of 1997. The Kyoto Protocol sets binding greenhouse gas emissions targets for countries that sign and ratify the agreement. The gases covered under the Protocol include carbon dioxide, methane, nitrous oxide, hydrofluorocarbons (HFCs), perfluorocarbons (PFCs) and sulfur hexafluoride.

lamp

A term generally used to describe artificial light. The term is often used when referring to a "bulb" or "tube."

land use

The ultimate uses to be permitted for currently contaminated lands, waters, and structures at each Department of Energy installation. Land-use decisions will strongly influence the cost of environmental management.

landfill gas

Gas that is generated by decomposition of organic material at landfill disposal sites. The average composition of landfill gas is approximately 50 percent methane and 50 percent carbon dioxide and water vapor by volume. The methane percentage, however, can vary from 40 to 60 percent, depending on several factors including waste composition (e.g. carbohydrate and cellulose content). The methane in landfill gas may be vented, flared, combusted to generate electricity or useful thermal energy on-site, or injected into a pipeline for combustion off-site.

land-use restrictions

Constraints placed upon mining by societal policies to protect surface features or entities that could be affected by mining. Because laws and regulations may be modified or repealed, the restrictions, including industrial and environmental restrictions, are subject to change.

- **langley**
  A unit or measure of solar radiation; 1 calorie per square centimeter or 3.69 Btu per square foot.

- **large passenger car**
  A passenger car with more than 120 cubic feet of interior passenger and luggage volume.

- **large pickup truck**
  A pickup truck weighing between 4,500-8,500 lbs gross vehicle weight (GVW).

- **latitude and longitude**
  The distance on the earth's surface measured, respectively, north or south of the equator and east or west of the standard meridian, expressed in angular degrees, minutes, and seconds.

- **leachate**
  The liquid that has percolated through the soil or other medium.

- **lead acid battery**
  An electrochemical battery that uses lead and lead oxide for electrodes and sulfuric acid for the electrolyte.

- **leaded gasoline**
  A fuel that contains more than 0.05 gram of lead per gallon or more than 0.005 gram of phosphorus per gallon.

- **leaded premium gasoline**
  Gasoline having an antiknock index (R+M/2) greater than 90 and containing more than 0.05 grams of lead or 0.005 grams of phosphorus per gallon.

- **leaded regular gasoline**
  Gasoline having an antiknock index (R+M/2) greater than or equal to 87 and less than or equal to 90 and containing more than 0.05 grams of lead or 0.005 grams of phosphorus per gallon.

- **leading edge**
  In reference to a wind energy conversion system, the area of a turbine blade surface that first comes into contact with the wind.

- **lease and plant fuel**
  Natural gas used in well, field, and lease operations (such as gas used in drilling operations, heaters, dehydrators, and field compressors) and as fuel in natural gas processing plants.

- **lease condensate**
  A mixture consisting primarily of pentanes and heavier hydrocarbons which is recovered as a liquid from natural gas in lease separation facilities. This category excludes natural gas plant liquids, such as butane and propane, which are recovered at downstream natural gas processing plants or facilities.

- **lease equipment**

All equipment located on the lease except the well to the point of the "Christmas tree."

- **lease fuel**

Natural gas used in well, field, and lease operations, such as gas used in drilling operations, heaters, dehydrators, and field compressors.

- **lease operations**

Any well, lease, or field operations related to the exploration for or production of natural gas prior to delivery for processing or transportation out of the field. Gas used in lease operations includes usage such as for drilling operations, heaters, dehydraters, field compressors, and net used for gas lift.

- **lease separation facility (lease separator)**

A facility installed at the surface for the purpose of (a) separating gases from produced crude oil and water at the temperature and pressure conditions set by the separator and/or (b) separating gases from that portion of the produced natural gas stream that liquefies at the temperature and pressure conditions set by the separator.

- **leasehold reserves**

Natural gas liquid reserves corresponding to the leasehold production defined above.

- **lessee**

An independent marketer who leases the station and land and has use of tanks, pumps, signs, etc. A lessee dealer typically has a supply agreement with a refiner or distributor and purchases products at dealer tank-wagon prices. The term "lessee dealer" is limited to those dealers who are supplied directly by a refiner or any affiliate or subsidiary of the reporting company. "Direct supply" includes use of commission agent or common carrier delivery.

- **levelized cost**

The present value of the total cost of building and operating a generating plant over its economic life, converted to equal annual payments. Costs are levelized in real dollars (i.e., adjusted to remove the impact of inflation).

- **leverage ratio**

A measure that indicates the financial ability to meet debt service requirements and increase the value of the investment to the stockholders. (i.e. the ratio of total debt to total assets).

- **liability**

An amount payable in dollars or by future services to be rendered.

- **licensed site capacity**

Capacity (number of assemblies) for which the site is currently licensed.

- **licensees**
Entity that has been granted permission to engage in an activity otherwise unlawful (i.e., hydropower project).

- **life extension**
Restoration or refurbishment of a plant to its original performance without the installation of new combustion technologies. Life extension results in 10 to 20 years of plant life beyond the anticipated retirement date, but usually does not result in larger capacity.

- **lift**
The force that pulls a wind turbine blade, as opposed to drag.

- **lifting costs**
The costs associated with the extraction of a mineral reserve from a producing property.

- **light bulbs**
A term generally used to describe a manmade source of light. The term is often used when referring to a "bulb" or "tube".

- **light gas oils**
Liquid petroleum distillates heavier than naphtha, with an approximate boiling range from 401 degrees to 650 degrees Fahrenheit.

- **light oil**
Lighter fuel oils distilled off during the refining process. Virtually all petroleum used in internal combustion and gas-turbine engines is light oil. Includes fuel oil numbers 1 and 2, kerosene, and jet fuel.

- **light rail**
An electric railway with a "light volume" traffic capacity compared to "heavy rail." Light rail may use exclusive or shared rights-of-way, high or low platform loading, and multi-car trains or single cars. Also known as "street car," "trolley car," and "tramway."

- **light trucks**
All single unit two-axle, four-tire trucks, including pickup trucks, sports utility vehicles, vans, motor homes, etc. This is the Department of Transportation definition. The Energy Information defined light truck as all trucks weighing 8,500 pounds or less.

- **light water**
Ordinary water (H_2O), as distinguished from heavy water or deuterium oxide (D_2O).

- **light water reactor (lwr)**
A nuclear reactor that uses water as the primary coolant and moderator, with slightly enriched uranium as fuel.

- **light-duty vehicles**
Vehicles weighing less than 8,500 lbs (include automobiles, motorcycles, and light trucks).

- **lighting conservation feature**
A building feature or practice designed to reduce the amount of

energy consumed by the lighting system.

- **lighting demand-side management (dsm) program**
  A DSM program designed to promote efficient lighting systems in new construction or existing facilities. Lighting DSM programs can include: certain types of high-efficiency fluorescent fixtures including T-8 lamp technology, solid state electronic ballasts, specular reflectors, compact fluorescent fixtures, LED and electro-luminescent Emergency Exist Signs, High Pressure Sodium with switchable ballasts, Compact Metal Halide, occupancy sensors, and day lighting controllers.

- **lighting equipment**
  These are light bulbs used to light the building's interior, such as incandescent light bulbs, fluorescent light bulbs, compact fluorescent light bulbs, and high-intensity discharge (HID) lights.

- **lights**
  All of the light bulbs controlled by one switch are counted as one light. For example, a chandelier with multiple lights controlled by one switch is counted as one light. A floor lamp with two separate globes or bulbs controlled by two separate switches would be counted as two lights. Indoor and outdoor lights were counted if they were under the control of the householder. This would exclude lights in the hallway of multifamily buildings.

- **lignite**
  The lowest rank of coal, often referred to as brown coal, used almost exclusively as fuel for steam-electric power generation. It is brownish-black and has a high inherent moisture content, sometimes as high as 45 percent The heat content of lignite ranges from 9 to 17 million Btu per ton on a moist, mineral-matter-free basis. The heat content of lignite consumed in the United States averages 13 million Btu per ton, on the as-received basis (i.e., containing both inherent moisture and mineral matter).

- **line loss**
  Electric energy lost because of the transmission of electricity. Much of the loss is thermal in nature.

- **line-miles of seismic exploration**
  The distance along the Earth's surface that is covered by seismic surveying.

- **liquefied natural gas (lng)**
  Natural gas (primarily methane) that has been liquefied by reducing its temperature to -260 degrees Fahrenheit at atmospheric pressure.

- **liquefied petroleum gases**
  A group of hydrocarbon-based gases derived from crude oil refining or natural gas fractionation. They include

ethane, ethylene, propane, propylene, normal butane, butylene, isobutane, and isobutylene. For convenience of transportation, these gases are liquefied through pressurization.

- **liquefied refinery gases (lrg)**
  Liquefied petroleum gases fractionated from refinery or still gases. Through compression and/ or refrigeration, they are retained in the liquid state. The reported categories are ethane/ethylene, propane/propylene, normal butane/butylene, and isobutane/ isobutylene.

- **liquid collector**
  A medium-temperature solar thermal collector, employed predominantly in water heating, which uses pumped liquid as the heat-transfer medium.

- **liquid metal fast breeder reactor**
  A nuclear breeder reactor, cooled by molten sodium, in which fission is caused by fast neutrons.

- **load (electric)**
  The amount of electric power delivered or required at any specific point or points on a system. The requirement originates at the energy-consuming equipment of the consumers.

- **load control program**
  A program in which the utility company offers a lower rate in return for having permission to turn off the air conditioner or water heater for short periods of time by remote control. This control allows the utility to reduce peak demand.

- **load curve**
  The relationship of power supplied to the time of occurrence. Illustrates the varying magnitude of the load during the period covered.

- **load diversity**
  The difference between the peak of coincident and noncoincident demands of two or more individual loads.

- **load factor**
  The ratio of the average load to peak load during a specified time interval.

- **load following**
  Regulation of the power output of electric generators within a prescribed area in response to changes in system frequency, tie line loading, or the relation of these to each other, so as to maintain the scheduled system frequency and/or established interchange with other areas within predetermined limits.

- **load leveling**
  Any load control technique that dampens the cyclical daily load flows and increases base load generation. Peak load pricing and

time-of-day charges are two techniques that electric utilities use to reduce peak load and to maximize efficient generation of electricity.

- **load loss (3 hours)**
  Any significant incident on an electric utility system that results in a continuous outage of 3 hours or longer to more than 50,000 customers or more than one half of the total customers being served immediately prior to the incident, whichever is less.

- **load management technique**
  Utility demand management practices directed at reducing the maximum kilowatt demand on an electric system and/or modifying the coincident peak demand of one or more classes of service to better meet the utility system capability for a given hour, day, week, season, or year.

- **load on equipment**
  One hundred percent load is the maximum continuous net output of the unit at normal operating conditions during the annual peak load month. For example, if the equipment is capable of operating at 5% overpressure continuously, use this condition for 100% load.

- **load reduction request**
  The issuance of any public or private request to any customer or the general public to reduce the use of electricity for the reasons of maintaining the continuity of service of the reporting entity's bulk electric power supply system. Requests to a customer(s) served under provisions of an interruptible contract are not a reportable action unless the request is made for reasons of maintaining the continuity of service of the reporting entity's bulk electric power supply.

- **load shape**
  A method of describing peak load demand and the relationship of power supplied to the time of occurrence.

- **load shedding**
  Intentional action by a utility that results in the reduction of more than 100 megawatts (MW) of firm customer load for reasons of maintaining the continuity of service of the reporting entity's bulk electric power supply system. The routine use of load control equipment that reduces firm customer load is not considered to be a reportable action.

- **local distribution company (ldc)**
  A legal entity engaged primarily in the retail sale and/or delivery of natural gas through a distribution system that includes mainlines (that is, pipelines designed to carry large volumes of gas, usually located under

roads or other major right-of-ways) and laterals (that is, pipelines of smaller diameter that connect the end user to the mainline). Since the restructuring of the gas industry, the sale of gas and/or delivery arrangements may be handled by other agents, such as producers, brokers, and marketers that are referred to as "non-LDC."

- **long ton**

A unit that equals 20 long hundredweight or 2,240 pounds. Used mainly in England.

- **long wall mining**

An automated form of underground coal mining characterized by high recovery and extraction rates, feasible only in relatively flat-lying, thick, and uniform coalbeds. A high-powered cutting machine is passed across the exposed face of coal, shearing away broken coal, which is continuously hauled away by a floor-level conveyor system. Long wall mining extracts all machine-minable coal between the floor and ceiling within a contiguous block of coal, known as a panel, leaving no support pillars within the panel area. Panel dimensions vary over time and with mining conditions but currently average about 900 feet wide (coal face width) and more than 8,000 feet long (the minable extent of the panel, measured in direction of mining). Longwall mining is done under movable roof supports that are advanced as the bed is cut. The roof in the mined-out area is allowed to fall as the mining advances.

- **long-term debt**

Debt securities or borrowings having a maturity of more than one year.

- **long-term purchase**

A purchase contract under which at least one delivery of material is scheduled to occur during the second calendar year after the contract-signing year. Deliveries also can occur during the contract-signing year, during the first calendar year thereafter, or during any subsequent calendar year.

- **loop flow**

The movement of electric power from generator to load by dividing along multiple parallel paths; it especially refers to power flow along an unintended path that loops away from the most direct geographic path or contract path.

- **loss of service (15 minutes)**

Any loss in service for greater than 15 minutes by an electric utility of firm loads totaling more than 200 MW, or 50 percent of the total load being supplied immediately prior to the incident, whichever is less. However, utilities with a peak load in the prior year of more than 3000 MW are only to report losses of service to firm loads totaling more than 300 MW for greater than 15 minutes. (The DOE shall be notified with service restoration and in any event, within three hours after the beginning of the interruption.)

- **low btu gas**

A fuel gas with a heating value between 90 and 200 Btu per cubic foot.

- **low e glass**

Low-emission glass reflects up to 90% of long-wave radiation, which is heat, but lets in short-wave radiation, which is light. Windows are glazed with a coating that bonds a microscopic, transparent, metallic substance to the inside surface of the double-pane or triple-pane windows.

- **low flow showerheads**

Reduce the amount of water flow through the showerhead from 5 to 6 gallons a minute to 3 gallons a minute.

- **low flush toilet**

A toilet that uses less water than a standard one during flushing, for the purpose of conserving water resources.

- **low head**

Vertical difference of 100 feet or less in the upstream surface water elevation (headwater) and the downstream surface water elevation (tailwater) at a dam.

- **low income home energy assistance program (liheap)**

The purpose of LIHEAP is to assist eligible households to meet the cost of heating or cooling in residential dwellings. The Federal government provides the funds to the States that administer the program.

- **low power testing**

The period of time between a plant's nuclear generating unit's initial fuel loading date and the issuance of its operating (full-power) license. The maximum level of operation during this period is 5 percent of the unit's thermal rating.

- **low temperature collectors**

Metallic or nonmetallic collectors that generally operate at temperatures below 110 degrees Fahrenheit and use pumped liquid or air as the heat transfer medium. They usually contain no glazing and no insulation, and they are often made of plastic or rubber, although some are made of metal.

- **low-pressure sodium lamp**

A type of lamp that produces light from sodium gas contained in a bulb operating at a partial pressure of 0.13 to 1.3 Pascal. The yellow light and large size make them applicable to lighting streets and parking lots.

lubricants
Substances used to reduce friction between bearing surfaces, or incorporated into other materials used as processing aids in the manufacture of other products, or used as carriers of other materials. Petroleum lubricants may be produced either from distillates or residues. Lubricants include all grades of lubricating oils, from spindle oil to cylinder oil to those used in greases.

lumen
An empirical measure of the quantity of light. It is based upon the spectral sensitivity of the photosensors in the human eye under high (daytime) light levels. Photometrically it is the luminous flux emitted with a solid angle (1 steradian) by a point source having a uniform luminous intensity of 1 candela.

lumens/watt (lpw)
A measure of the efficacy (efficiency) of lamps. It indicates the amount of light (lumens) emitted by the lamp for each unit of electrical power (Watts) used.

machine drive (motors)
The direct process end use in which thermal or electric energy is converted into mechanical energy. Motors are found in almost every process in manufacturing. Therefore, when motors are found in equipment that is wholly contained in another end use (such as process cooling and refrigeration), the energy is classified there rather than in machine drive.

made available (vehicle)
A vehicle is considered "Made available" if it is available for delivery to dealers or users, whether or not it was actually delivered to them. To be "Made available", the vehicle must be completed and available for delivery; thus, any conversion to be performed by an original equipment manufacturer (OEM) Vehicle Converter or Aftermarket Vehicle Converter must have been completed.

magma
Naturally occurring molten rock, generated within the earth and capable of intrusion and extrusion, from which igneous rocks are thought to have been derived through solidification and related processes. It may or may not contain suspended solids (such as crystals and rock fragments) and/or gas phases.

main heating equipment
Equipment primarily used for heating ambient air in the housing unit.

main heating fuel
The form of energy used most frequently to heat the largest portion of the floor space of a structure. The energy source

designated as the main heating fuel is the source delivered to the site for that purpose, not any subsequent form into which it is transformed on site to deliver the heat energy (e.g., for buildings heated by a steam boiler, the main heating fuel is the main input fuel to the boiler, not the steam or hot water circulated through the building.) Note: In commercial buildings, the heating must be to at least 50 degrees Fahrenheit.

- **mains**
  A system of pipes for transporting gas within a distributing gas utility's retail service area to points of connection with consumer service pipes.

- **maintenance expenses**
  That portion of operating expenses consisting of labor, materials, and other direct and indirect expenses incurred for preserving the operating efficiency and/or physical condition of utility plants used for power production, transmission, and distribution of energy.

- **maintenance of boiler plant (expenses)**
  The cost of labor, material, and expenses incurred in the maintenance of a steam plant. Includes furnaces; boilers; coal, ash-handling, and coal-preparation equipment; steam and feed water piping; and boiler apparatus and accessories used in the production of steam, mercury, or other vapor to be used primarily for generating electricity. The point at which an electric steam plant is distinguished from an electric plant is defined as follows: 1. Inlet flange of throttle valve on prime mover. 2. Flange of all steam extraction lines on prime mover. 3. Hotwell pump outlet on condensate lines. 4. Inlet flange of all turbine-room auxiliaries. 5. Connection to line side of motor starter for all boiler-plant equipment.

- **maintenance of structures (expenses)**
  The cost of labor, materials, and expenses incurred in maintenance of power production structures. Structures include all buildings and facilities to house, support, or safeguard property or persons.

- **maintenance supervision and engineering expenses**
  The cost of labor and expenses incurred in the general supervision and direction of the maintenance of power generation stations. The supervision and engineering included consists of the pay and expenses of superintendents, engineers, clerks, other employees, and consultants engaged in supervising and directing the maintenance of each utility function. Direct supervision and engineering of specific activities,

such as fuel handling, boiler room operations, generator operations, etc., are charged to the appropriate accounts.

- **major electric utility**

A utility that, in the last 3 consecutive calendar years, had sales or transmission services exceeding one of the following: (1) 1 million megawatt hours of total annual sales; (2) 100 megawatt hours of annual sales for resale; (3) 500 megawatt hours of annual gross interchange out; or (4) 500 megawatt hours of wheeling (deliveries plus losses) for others.

- **major energy sources**

Fuels or energy sources such as electricity, fuel oil, natural gas, district steam, district hot water, and district chilled water. District chilled water is not included in any totals for the sum of major energy sources or fuels; all other major fuels are included in these totals.

- **major fuels**

Fuels or energy sources such as: electricity, fuel oil, liquefied petroleum gases, natural gas, district steam, district hot water, and district chilled water.

- **major interstate pipeline company**

A company whose combined sales for resale, including gas transported interstate or stored for a fee, exceeded 50 million thousand cubic feet in the previous year.

- **make-up air**

Air brought into a building from outside to replace exhaust air.

- **manhattan project**

The U.S. Government project that produced the first nuclear weapons during World War II. Started in 1942, the Manhattan Project formally ended in 1946. The Hanford Site, Oak Ridge Reservation, and Los Alamos National Laboratory were created for this effort. The project was named for the Manhattan Engineer District of the U.S. Army Corps of Engineers.

- **manual dimmer switches**

These are like residential-style dimmer switches. They are not generally used with fluorescent and high-intensity discharge (HID) lamps.

- **manufactured gas**

A gas obtained by destructive distillation of coal or by the thermal decomposition of oil, or by the reaction of steam passing through a bed of heated coal or coke. Examples are coal gases, coke oven gases, producer gas, blast furnace gas, blue (water) gas, carbureted water gas. Btu content varies widely.

- **manufacturing**

An energy-consuming subsector of the industrial sector that consists of all facilities and equipment engaged in the mechanical, physical, chemical, or electronic transformation of materials, substances, or components into new products. Assembly of component parts of products is included, except for that which is included in construction.

- **manufacturing division**
One of 10 fields of economic activity defined by the Standard Industrial Classification Manual. The manufacturing division includes all establishments engaged in the mechanical or chemical transformation of materials or substances into new products. The other divisions of the U.S. economy are agriculture, forestry, fishing, hunting, and trapping; mining; construction; transportation, communications, electric, gas, and sanitary services; wholesale trade; retail trade; finance, insurance, and real estate; personal, business, professional, repair, recreation, and other services; and public administration. The establishments in the manufacturing division constitute the universe for the MECS.

- **manufacturing establishment**
An economic unit at a single physical location where mechanical or chemical transformation of materials or substances into new products are performed.

- **marginal cost**
The change in cost associated with a unit change in quantity supplied or produced.

- **marine freight**
Freight transported over rivers, canals, the Great Lakes, and domestic ocean waterways.

- **market clearing price**
The price at which supply equals demand for the Day-ahead or hour-ahead markets.

- **market price contract**
A contract in which the price of uranium is not specifically determined at the time the contract is signed but is based instead on the prevailing market price at the time of delivery. A market price contract may include a floor price, that is, a lower limit on the eventual settled price. The floor price and the method of price escalation generally are determined when the contract is signed. The contract may also include a price ceiling or a discount from the agreed-upon market price reference.

- **market price settlement (uranium)**
The price paid for uranium delivery under a market-price contract. The price is commonly (but not always) determined at or sometime before delivery and may be related to a floor price, ceiling price, or discount.

marketable coke

Those grades of coke produced in delayed or fluid cokers that may be recovered as relatively pure carbon. This "green" coke may be sold as is or further purified by calcining.

market-based pricing

Prices of electric power or other forms of energy determined in an open market system of supply and demand under which prices are set solely by agreement as to what buyers will pay and sellers will accept. Such prices could recover less or more than full costs, depending upon what the buyers and sellers see as their relevant opportunities and risks.

marketed production

Gross withdrawals less gas used for repressuring, quantities vented and flared, and nonhydrocarbon gases removed in treating or processing operations. Includes all quantities of gas used in field and processing plant operations.

masonry

A general term covering wall construction using masonry materials such as brick, concrete block, stone, and tile that are set in mortar; also included is stucco. The category does not include concrete panels because concrete panels represent a different method of constructing buildings. Concrete panels are reported separately.

masonry stove

A type of heating appliance similar to a fireplace, but much more efficient and clean burning. They are made of masonry and have long channels through which combustion gases give up their heat to the heavy mass of the stove, which releases the heat slowly into a room. Often called Russian or Finnish fireplaces.

mass burn facility

A type of municipal solid waste (MSW) incineration facility in which MSW is burned with only minor presorting to remove oversize, hazardous, or explosive materials.

master-metering

Measurement of electricity or natural gas consumption of several tenants or housing units using a single meter. That is, one meter measures the energy usage for several households collectively.

maximum deliverability

The maximum deliverability rate (Mcf/d) estimated at the present developed maximum operating capacity.

maximum demand

The greatest of all demands of the load that has occurred within a specified period of time.

- **maximum dependable capacity, net**

The gross electrical output measured at the output terminals of the turbine generator(s) during the most restrictive seasonal conditions, less the station service load.

- **maximum established site capacity (reactors)**

The maximum established spent fuel capacity for the site is defined by DOE as the maximum number of intact assemblies that will be able to be stored at some point in the future (between the reporting date and the reactor's end of life) taking into account any established or current studies or engineering evaluations at the time of submittal for licensing approval from the NRC.

- **maximum hourly load**

This is determined by the interval in which the 60-minute integrated demand is the greatest.

- **maximum stream flow**

The maximum rate of water flow past a given point during a specified period.

- **mcf**

One thousand cubic feet.

- **mean indoor temperature**

The "usual" temperature. If different sections of the house are kept at different temperatures, the reported temperature is for the section where the people are. A thermostat setting is accepted if the temperature is not known.

- **mean operating hours**

The arithmetic average number of operating hours per building is the weighted sum of the number of operating hours divided by the weighted sum of the number of buildings.

- **mean power output (of a wind turbine)**

The average power output of a wind energy conversion system at a given mean wind speed based on a Raleigh frequency distribution.

- **mean square feet per building**

The arithmetic average square feet per building is the weighted sum of the total square feet divided by the weighted sum of the number of buildings.

- **measured heated area of residence**

The floor area of the housing unit that is enclosed from the weather and heated. Basements are included whether or not they contain finished space. Garages are included if they have a wall in common with the house. Attics that have finished space and attics that have some heated space are included. Crawl spaces are not included even if they are enclosed from the weather. Sheds and other buildings that are not attached to the house are not included.

"Measured" area means the measurement of the dimensions of the home, using a metallic, retractable, 50-foot tape measure. "Heated area" is that portion of the measured area that is heated during most of the season. Rooms that are shut off during the heating season to save on fuel are not counted. Attached garages that are unheated and unheated areas in the attics and basements are also not counted.

- **measured resources, coal**

Coal resources for which estimates of the rank, quality, and quantity have been computed, within a margin of error of less than 20 percent, from sample analyses and measurements from closely spaced and geologically well known sample sites. Measured resources are computed from dimensions revealed in outcrops, trenches, mine workings, and drill holes. The points of observation and measurement are so closely spaced and the thickness and extent of coals are so well defined that the tonnage is judged to be accurate within 20 percent. Although the spacing of the points of observation necessary to demonstrate continuity of the coal differs from region to region, according to the character of the coalbeds, the point of observation are no greater than 1/2 mile apart. Measured coal is projected to extend as a belt 1/4 mile wide from the outcrop or points of observation or measurement.

- **median stream flow**

The middle rate of flow of water past a given point for which there have been several greater and lesser rates of flow occurring during a specified period.

- **median**

The middle number of a data set when the measurements are arranged in ascending (or descending) order.

- **median water condition**

The middle precipitation and run-off condition for a distribution of water conditions that have happened over a long period of time. Usually determined by examining the water supply record of the period in question.

- **medium pressure**

For valves and fittings, implies that they are suitable for working pressures between 125 to 175 pounds per square inch.

- **medium-temperature collector**

A collector designed to operate in the temperature range of 140 degrees to 180 degrees Fahrenheit, but that can also operate at a temperature as low as 110 degrees Fahrenheit. The collector typically consists of a metal frame, metal absorption panels with integral flow channels (attached tubing for liquid collectors or integral ducting for air collectors), and glazing and insulation on the sides and back.

■ **megavoltamperes (mva)**
Millions of voltamperes, which are a measure of apparent power.

■ **megawatt (mw)**
One million watts of electricity.

■ **megawatt electric (mwe)**
One million watts of electric capacity.

■ **megawatthour (mwh)**
One thousand kilowatt-hours or 1 million watt-hours.

■ **mercaptan**
An organic chemical compound that has a sulfur like odor that is added to natural gas before distribution to the consumer, to give it a distinct, unpleasant odor (smells like rotten eggs). This serves as a safety device by allowing it to be detected in the atmosphere, in cases where leaks occur.

■ **merchant coke plant**
A coke plant where coke is produced primarily for sale on the commercial (open) market.

■ **merchant facilities**
High-risk, high-profit facilities that operate, at least partially, at the whims of the market, as opposed to those facilities that are constructed with close cooperation of municipalities and have significant amounts of waste supply guaranteed.

■ **merchant mtbe plants**
MTBE (methyl tertiary butyl ether) production facilities primarily located within petrochemical plants rather than refineries. Production from these units is sold under contract or on the spot market to refiners or other gasoline blenders.

■ **merchant oxygenate plants**
Oxygenate production facilities that are not associated with a petroleum refinery. Production from these facilities is sold under contract or on the spot market to refiners or other gasoline blenders.

■ **mercury vapor lamp**
A high-intensity discharge lamp that uses mercury as the primary light-producing element. Includes clear, phosphor coated, and self-ballasted lamps.

■ **merger**
A combining of companies or corporations into one, often by issuing stock of the controlling corporation to replace the greater part of that of the other.

■ **met**
An approximate unit of heat produced by a resting person, equal to about 18.5 Btu per square foot per hour.

- **metal halide lamp**
  A high-intensity discharge lamp type that uses mercury and several halide additives as light-producing elements. These lights have the best Color Rendition Index (CRI) of the high-intensity discharge lamps. They can be used for commercial interior lighting or for stadium lights.

- **metallic**
  The metallic material composition of the collector's absorber system.

- **metallurgical coal**
  Coking coal and pulverized coal consumed in making steel.

- **metered data**
  End-use data obtained through the direct measurement of the total energy consumed for specific uses within the individual household. Individual appliances can be submetered by connecting the recording meters directly to individual appliances.

- **metered peak demand**
  The presence of a device to measure the maximum rate of electricity consumption per unit of time. This device allows electric utility companies to bill their customers for maximum consumption, as well as for total consumption.

- **methane**
  A colorless, flammable, odorless hydrocarbon gas (CH_4) which is the major component of natural gas. It is also an important source of hydrogen in various industrial processes. Methane is a greenhouse gas.

- **methanogens**
  Bacteria that synthesize methane, requiring completely anaerobic conditions for growth.

- **methanol (ch3oh)**
  A light, volatile alcohol eligible for gasoline blending.

- **methanol blend**
  Mixtures containing 85 percent or more (or such other percentage, but not less than 70 percent) by volume of methanol with gasoline. Pure methanol is considered an "other alternative fuel."

- **methanotrophs**
  Bacteria that use methane as food and oxidize it into carbon dioxide.

- **methyl chloroform (trichloroethane)**
  An industrial chemical (CH_3CCl_3) used as a solvent, aerosol propellant, and pesticide and for metal degreasing.

- **methylene chloride**
  A colorless liquid, nonexplosive and practically nonflammable. Used as a refrigerant in centrifugal compressors, a solvent for organic materials, and a component in nonflammable paint removers.

■ **metric conversion factors (for floorspace)**

Floorspace estimates may be converted to metric units by using the relationship, 1 square foot is approximately equal to .0929 square meters. Energy estimates may be converted to metric units by using the relationship, 1 Btu is approximately equal to 1,055 joules. One kilowatthour is exactly 3,600,000 joules. One gigajoule is approximately 278 kilowatthours (kWh).

■ **metric ton**

A unit of weight equal to 2,204.6 pounds.

■ **metropolitan area**

A geographic area that is a metropolitan statistical area or a consolidated metropolitan statistical area as defined by the U.S. Office of Management and Budget.

■ **metropolitan**

Located within the boundaries of a metropolitan area.

■ **metropolitan statistical area**

A county or group of contiguous counties (towns and cities in New England) that has (1) at least one city with 50,000 or more inhabitants; or (2) an urbanized area of 50,000 inhabitants and a total population of 100,000 or more inhabitants (75,000 in New England). These areas are defined by the U.S. Office of Management and Budget. The contiguous counties or other jurisdictions to be included in an MSA are those that, according to certain criteria, are essentially metropolitan in character and are socially and economically integrated with the central city or urbanized area.

■ **microcrystalline wax**

Wax extracted from certain petroleum residues having a finer and less apparent crystalline structure than paraffin wax and having the following physical characteristics: penetration at 77 degrees Fahrenheit (D1321)-60 maximum; viscosity at 210 degrees Fahrenheit in Saybolt Universal Seconds (SUS); (D88)-60 SUS (10.22 centistokes) minimum to 150 SUS (31.8 centistokes) maximum; oil content (D721)-5 percent minirnum.

■ **microgroove**

A small groove scribed into the surface of a solar photovoltaic cell which is filled with metal for contacts.

■ **micrometer (or micron)**

One-millionth of a meter. It can also be expressed as 10^{-6} meter.

■ **microwave oven**

A household cooking appliance consisting of a compartment designed to cook or heat food by means of microwave energy. It may also have a browning coil and convection heating as additional features.

middle distillates
A general classification of refined petroleum products that includes distillate fuel oil and kerosene.

middlings
In coal preparation, this material called mid-coal is neither clean nor refuse; due to their intermediate specific gravity, middlings sink only partway in the washing vessels and are removed by auxiliary means.

midgrade gasoline
Gasoline having an antiknock index, i.e., octane rating, greater than or equal to 88 and less than or equal to 90. Note: Octane requirements may vary by altitude.

mid-size passenger car
A passenger car with between 110 and 119 cubic feet of interior passenger and luggage volume.

miles per gallon (mpg)
A measure of vehicle fuel efficiency. Miles per gallon or MPG represents "Fleet Miles per Gallon." For each subgroup or "table cell," MPG is computed as the ratio of the total number of miles traveled by all vehicles in the subgroup to the total number of gallons consumed. MPGs are assigned to each vehicle using the EPA certification files and adjusted for on-road driving.

military use
Includes sales to the Armed Forces, including volumes sold to the Defense Fuel Supply Center (DFSC) for use by all branches of the Department of Defense (DOD).

mill
A monetary cost and billing unit used by utilities; it is equal to 1/1000 of the U.S. dollar (equivalent to 1/10 of 1 cent).

mill capital
Cost for transportation and equipping a plant for processing ore or other feed materials.

mill feed
Uranium ore supplied to a crusher or grinding mill in an ore-dressing process.

milling capacity
The maximum rate at which a mill is capable of treating ore or producing concentrate.

milling of uranium
The processing of uranium from ore mined by conventional methods, such as underground or openpit methods, to separate the uranium from the undesired material in the ore.

milling
The grinding or crushing of ore, concentration, and other benefication, including the removal of valueless or harmful constituents and preparation for market.

minable
Capable of being mined under current mining technology and environmental and legal restrictions, rules, and regulations.

mine capital
Cost for exploration and development, pre-mining stripping, shaft sinking, and mine development (including in situ leaching), as well as the mine plant and its equipment.

mine count
The number of mines, or mines collocated with preparation plants or tipples, located in a particular geographic area (state or region). If a mine is mining coal across two counties within a state, or across two states, then it is counted as two operations. This is done so that EIA can separate production by state and county.

mineral
Any of the various naturally occurring inorganic substances, such as metals, salt, sand, stone, sulfur, and water, usually obtained from the earth. Note: For reporting on the Financial Reporting System the term also includes organic non-renewable substances that are extracted from the earth such as coal, crude oil, and natural gas.

mineral lease
An agreement wherein a mineral interest owner (lessor) conveys to another party (lessee) the rights to explore for, develop, and produce specified minerals. The lessee acquires a working interest and the lessor retains a non-operating interest in the property, referred to as the royalty interest, each in proportions agreed upon.

mineral rights
The ownership of the minerals beneath the earth's surface with the right to remove them. Mineral rights may be conveyed separately from surface rights.

mineral-matter-free basis
Mineral matter in coal is the parent material in coal from which ash is derived and which comes from minerals present in the original plant materials that formed the coal, or from extraneous sources such as sediments and precipitates from mineralized water. Mineral matter in coal cannot be analytically determined and is commonly calculated using data on ash and ash-forming constituents. Coal analyses are calculated to the mineral matter free basis by adjusting formulas used in calculations in order to deduct the weight of mineral matter from the total coal.

mini van
Small van that first appeared with that designation in 1984. Any of the smaller vans built on an

automobile-type frame. Earlier models such as the Volkswagen van are now included in this category.

- **minimum stream flow**
The lowest rate of flow of water past a given point during a specified period.

- **mining**
An energy-consuming subsector of the industrial sector that consists of all facilities and equipment used to extract energy and mineral resources.

- **mining operation**
One mine and/or tipple at a single physical location.

- **minority carrier**
A current carrier, either an electron or a hole, that is in the minority in a specific layer of a semiconductor material; the diffusion of minority carriers under the action of the cell junction voltage is the current in a photovoltaic device.

- **miscellaneous petroleum products**
Includes all finished products not classified elsewhere (e.g., petrolatum lube refining byproducts (aromatic extracts and tars), absorption oils, ram-jet fuel, petroleum rocket fuels, synthetic natural gas feed stocks, and specialty oils).

- **miscellaneous reserves**
A supply source having not more than 50 billion cubic feet of dedicated recoverable salable reserves and that falls within the definition of **Supply Source**.

- **mixed waste**
Waste containing both radioactive and hazardous constituents.

- **mmbbl/d**
One million barrels of oil per day.

- **mmbtu**
One million British thermal units.

- **mmcf**
One million cubic feet.

- **mmgal/d**
One million gallons per day.

- **mmst**
One million short tons.

- **mobile home**
A housing unit built on a movable chassis and moved to the site. It may be placed on a permanent or temporary foundation and may contain one room or more. If rooms are added to the structure, it is considered a single-family housing unit. A manufactured house assembled on site is a single-family housing unit, not a mobile home.

- **moderator**
A material, such as ordinary water, heavy water, or graphite, used in a reactor to slow down high-velocity neutrons, thus

increasing the likelihood of further fission.

■ modules

Photovoltaic cells or an assembly of cells into panels (modules) intended for and shipped for final consumption or to another organization for resale. When exported, incomplete modules and unencapsulated cells are also included. Modules used for space applications are not included.

■ moist (coal) basis

"Moist" coal contains its natural inherent or bed moisture, but does not include water adhering to the surface. Coal analyses expressed on a moist basis are performed or adjusted so as to describe the data when the coal contains only that moisture that exists in the bed in its natural state of deposition and when the coal has not lost any moisture due to drying.

■ moisture content

The water content of a substance (a solid fuel) as measured under specified conditions being the "dry basis," which equals the weight of the wet sample minus the weight of a (bone) dry sample divided by the weight of the dry sample times 100 (to get percent); "wet basis," which is equal to the weight of the wet sample minus the weight of the dry sample divided by the weight of the wet sample times 100.

■ mole

The quantity of a compound or element that has a weight in grams numerically equal to its molecular weight. Also referred to as "gram molecule" or "gram molecular weight."

■ montreal protocol

The Montreal Protocol on Substances that Deplete the Ozone Layer. An international agreement, signed by most of the industrialized nations, to substantially reduce the use of chlorofluorocarbons (CFCs). Signed in January 1989, the original document called for a 50-percent reduction in CFC use by 1992 relative to 1986 levels. The subsequent London Agreement called for a complete elimination of CFC use by 2000. The Copenhagen Agreement, which called for a complete phase out by January 1, 1996, was implemented by the U.S. Environmental Protection Agency.

■ motor gasoline (finished)

A complex mixture of relatively volatile hydrocarbons with or without small quantities of additives, blended to form a fuel suitable for use in spark-ignition engines. Motor gasoline, as defined in ASTM Specification D 4814 or Federal Specification VV-G-1690C, is characterized as having a boiling range of 122 to 158 degrees Fahrenheit at the 10 percent recovery point to 365 to 374 degrees Fahrenheit at the 90 percent recovery point. Motor Gasoline includes conventional gasoline; all types of oxygenated gasoline, including gasohol; and reformulated gasoline, but excludes aviation gasoline.

- **motor gasoline blending components**

Naphthas (e.g., straight-run gasoline, alkylate, reformate, benzene, toluene, xylene) used for blending or compounding into finished motor gasoline. These components include reformulated gasoline blendstock for oxygenate blending (RBOB) but exclude oxygenates (alcohols, ethers), butane, and pentanes plus. Note: Oxygenates are reported as individual components and are included in the total for other hydrocarbons, hydrogen and oxygenates.

- **motor gasoline blending**

Mechanical mixing of motor gasoline blending components, and oxygenates when required, to produce finished motor gasoline. Finished motor gasoline may be further mixed with other motor gasoline blending components or oxygenates, resulting in increased volumes of finished motor gasoline and/or changes in the formulation of finished motor gasoline (e.g., conventional motor gasoline mixed with MTBE to produce oxygenated motor gasoline).

- **motor speed**

The number of revolutions that the motor turns in a given time period (i.e. revolutions per minute, rpm).

- **mpg**

Miles per gallon.

- **mpg shortfall**

The difference between actual on-road MPG and EPA laboratory test MPG. MPG shortfall is expressed as gallons per mile ratio (GPMR).

- **msha id number**

Seven (7)-digit code assigned to a mining operation by the Mine Safety and Health Administration.

- **mtbe (methyl tertiary butyl ether) (ch3)3coch3**

An ether intended for gasoline blending as described in "**Oxygenates.**"

- **multiple completion well**

A well equipped to produce oil and/or gas separately from more than one reservoir. Such wells contain multiple strings of tubing or other equipment that permit production from the various completions to be measured and accounted for separately. For statistical purposes, a multiple completion well is reported as one well and classified as either an oil well or a gas well. If one of the several completions in a given well is an oil completion, the well is classified as an oil well. If all of the completions in a given well are gas completions, the well is classified as a gas well.

- **multiple cropping**

A system of growing several crops on the same field in one year.

- **multiple purpose project**

The development of hydroelectric facilities to serve more than one function. Some of the uses include hydroelectric power, irrigation, water supply, water quality control, and/or fish and wildlife enhancement.

- **multiple purpose reservoir**

Stored water and its usage governed by advanced water resource conservation practices to achieve more than one water control objective. Some of the objectives include flood control, hydroelectric power development, irrigation, recreation usage, and wilderness protection.

- **municipal solid waste**

Residential solid waste and some nonhazardous commercial, institutional, and industrial wastes.

- **municipal waste**

As defined in the Energy Security Act as "any organic matter, including sewage, sewage sludge, and industrial or commercial waste, and mixtures of such matter and inorganic refuse from any publicly or privately operated municipal waste collection or similar disposal system, or from similar waste flows (other than such flows which constitute agricultural wastes or residues, or wood wastes or residues from wood harvesting activities or production of forest products)."

- **municipal waste to energy project or plant**

A facility that produces fuel or energy from municipal solid waste.

- **municipality**

A village town, city, county, or other political subdivision of a State.

- **naics**

A coding system developed jointly by the United States, Canada, and Mexico to classify businesses and industries according to the type of economic activity in which they are engaged. NAICS replaces the Standard Industrial Classification (SIC) codes.

- **name plate**

A metal tag attached to a machine or appliance that contains information such as brand name, serial number, voltage, power ratings under specified conditions, and other manufacturer supplied data.

- **naphtha**

A generic term applied to a petroleum fraction with an

approximate boiling range between 122 degrees Fahrenheit and 400 degrees Fahrenheit.

- **naphthas**

Refined or partly refined light distillates with an approximate boiling point range of 27 degrees to 221 degrees Centigrade. Blended further or mixed with other materials, they make high-grade motor gasoline or jet fuel. Also, used as solvents, petrochemical feed stocks, or as raw materials for the production of town gas.

- **naphtha-type jet fuel**

A fuel in the heavy naphtha boiling range having an average gravity of 52.8 degrees API, 20 to 90 percent distillation temperatures of 290 degrees to 470 degrees Fahrenheit, and meeting Military Specification MIL-T-5624L (Grade JP-4). It is used primarily for military turbojet and turboprop aircraft engines because it has a lower freeze point than other aviation fuels and meets engine requirements at high altitudes and speeds.

- **national association of regulatory utility commissioners (naruc)**

An affiliation of the public service commissioners to promote the uniform treatment of members of the railroad, public utilities, and public service commissions of the 50 states, the District of Columbia, the Commonwealth of Puerto Rico, and the territory of the Virgin Islands.

- **national defense authorization act**

The federal law, enacted in 1994 and amended in 1995, that required the Secretary of Energy to prepare the Baseline Report.

- **national priorities list**

The Environmental Protection Agency's list of the most serious uncontrolled or abandoned hazardous waste sites identified for possible long-term remedial action under the Comprehensive Environmental Response, Compensation, and Liability Act (CERCLA). The list is based primarily on the score a site receives from the Environmental Protection Agency Hazard Ranking System. The Environmental Protection Agency is required to update the National Priorities List at least once a year.

- **national rural electric cooperative association (nreca)**

A national organization dedicated to representing the interests of cooperative electric utilities and the consumers they serve. Members come from the 46 states that have an electric distribution cooperative.

- **national uranium resource evaluation (nure)**
A program begun by the U.S. Atomic Energy Commission (AEC) in 1974 to make a comprehensive evaluation of U.S. uranium resources and continued through 1983 by the AEC's successor agencies, the Energy Research and Development Administration (ERDA), and the Department of Energy (DOE). The NURE program included aerial radiometric and magnetic surveys, hydrogeo chemical and stream sediment surveys, geologic drilling in selected areas, geophysical logging of selected boreholes, and geologic studies to identify and evaluate geologic environments favorable for uranium.

- **native gas**
Gas in place at the time that a reservoir was converted to use as an underground storage reservoir in contrast to injected gas volumes.

- **natural gas**
A gaseous mixture of hydrocarbon compounds, the primary one being methane.

- **natural gas field facility**
A field facility designed to process natural gas produced from more than one lease for the purpose of recovering condensate from a stream of natural gas; however, some field facilities are designed to recover propane, normal butane, pentanes plus, etc., and to control the quality of natural gas to be marketed.

- **natural gas gross withdrawals**
Full well-stream volume of produced natural gas, excluding condensate separated at the lease.

- **natural gas hydrates**
Solid, crystalline, wax-like substances composed of water, methane, and usually a small amount of other gases, with the gases being trapped in the interstices of a water-ice lattice. They form beneath permafrost and on the ocean floor under conditions of moderately high pressure and at temperatures near the freezing point of water.

- **natural gas liquids (ngl)**
Those hydrocarbons in natural gas that are separated from the gas as liquids through the process of absorption, condensation, adsorption, or other methods in gas processing or cycling plants. Generally such liquids consist of propane and heavier hydrocarbons and are commonly referred to as lease condensate, natural gasoline, and liquefied petroleum gases. Natural gas liquids include natural gas plant liquids and lease condensate.

- **natural gas liquids production**
The volume of natural gas liquids removed from natural gas in lease

separators, field facilities, gas processing plants, or cycling plants during the report year.

- **natural gas marketed production**

Gross withdrawals of natural gas from production reservoirs, less gas used for reservoir repressuring, nonhydrocarbon gases removed in treating and processing operations, and quantities vented and flared.

- **natural gas plant liquids**

Those hydrocarbons in natural gas that are separated as liquids at natural gas processing plants, fractionating and cycling plants, and, in some instances, field facilities. Lease condensate is excluded. Products obtained include ethane; liquefied petroleum gases (propane, butanes, propane-butane mixtures, ethane-propane mixtures); isopentane; and other small quantities of finished products, such as motor gasoline, special naphthas, jet fuel, kerosene, and distillate fuel oil.

- **natural gas policy act of 1978 (ngpa)**

Signed into law on November 9, 1978, the NGPA is a framework for the regulation of most facets of the natural gas industry.

- **natural gas processing plant**

Facilities designed to recover natural gas liquids from a stream of natural gas that may or may not have passed through lease separators and/or field separation facilities. These facilities control the quality of the natural gas to be marketed. Cycling plants are classified as gas processing plants.

- **natural gas utility demand-side management (dsm) program sponsor**

A DSM (demand-side management) program sponsored by a natural gas utility that suggests ways to increase the energy efficiency of buildings, to reduce energy costs, to change the usage patterns, or to promote the use of a different energy source.

- **natural gasoline**

A term used in the gas processing industry to refer to a mixture of liquid hydrocarbons (mostly pentanes and heavier hydrocarbons) extracted from natural gas. It includes isopentane.

- **natural gasoline and isopentane**

A mixture of hydrocarbons, mostly pentanes and heavier, extracted from natural gas, that meets vapor pressure, end-point, and other specifications for natural gasoline set by the Gas Processors Association. Includes isopentane which is a saturated branch-chain hydrocarbon, (C_5H_{12}), obtained by fractionation of natural gasoline or isomerization of normal pentane.

natural reservoir pressure
The energy within an oil or gas reservoir that causes the oil or gas to rise (unassisted by other forces) to the earth's surface when the reservoir is penetrated by an oil or gas well. The energy may be the result of "dissolved gas drive," "gas cap drive," or "water drive." Regardless of the type of drive, the principle is the same: the energy of the gas or water, creating a natural pressure, forces the oil or gas to the well bore.

natural stream flow
The rate of flow of water past a given point of an uncontrolled stream or regulated stream flow adjusted to eliminate the effects of reservoir storage or upstream diversions at a set time interval.

natural uranium
Uranium with the U-235 isotope present at a concentration of 0.711 percent (by weight), that is, uranium with its isotopic content exactly as it is found in nature.

nerc
North American Electric Reliability Council. See definition further down on this page.

net cell shipments
Represents the difference between cell shipments and cell purchases.

net electricity consumption
Consumption of electricity computed as generation, plus imports, minus exports, minus transmission and distribution losses.

net energy for load
Net generation of main generating units that are system-owned or system-operated, plus energy receipts minus energy deliveries.

net energy for system
The sum of energy an electric utility needs to satisfy their service areas, including full and partial requirements consumers.

net generation
The amount of gross generation less the electrical energy consumed at the generating station(s) for station service or auxiliaries. Note: Electricity required for pumping at pumped-storage plants is regarded as electricity for station service and is deducted from gross generation.

net head
The gross head minus all hydraulic losses except those chargeable to the turbine.

net income
Operating income plus other income and extraordinary income less operating expenses, taxes, interest charges, other deductions, and extraordinary deductions.

net interstate flow of electricity
The difference between the sum of electricity sales and losses

within a state and the total amount of electricity generated within that state. A positive number indicates that more electricity (including associated losses) came into the state than went out of the state during the year; conversely, a negative number indicates that more electricity (including associated losses) went out of the state than came into the state.

- **net module shipments**
Represents the difference between module shipments and module purchases. When exported, incomplete modules and unencapsulated cells are also included.

- **net operable capacity**
Total owned capacity less inoperable capacity.

- **net photovoltaic module shipment**
The difference between photovoltaic module shipments and photovoltaic module purchases.

- **net profits interest**
A contractual arrangement under which the beneficiary, in exchange for consideration paid, receives a stated percentage of the net profits. That type of arrangement is considered a non-operating interest, as distinguished from a working interest, because it does not involve the rights and obligations of operating a mineral property (costs of exploration, development, and operation). The net profits interest does not bear any part of net losses.

- **net receipts**
The difference between total movements into and total movements out of each PAD District by pipeline, tanker, and barge.

- **net summer capacity**
The maximum output, commonly expressed in megawatts (MW), that generating equipment can supply to system load, as demonstrated by a multi-hour test, at the time of summer peak demand. This output reflects a reduction in capacity due to electricity use for station service or auxiliaries.

- **net winter capacity**
The maximum output, commonly expressed in megawatts (MW), that generating equipment can supply to system load, as demonstrated by a multi-hour test, at the time of peak winter demand (period of November 1 though April 30). This output reflects a reduction in capacity due to electricity use for station service or auxiliaries.

- **netback purchase**
Refers to a crude oil purchase agreement wherein the price paid

for the crude is determined by sales prices of the types of products that are derivable from that crude as well as other considerations (e.g., transportation and processing costs). Typically, the price is calculated based on product prices extant on or near the cargo's date of importation.

- **new field**
  A field discovered during the report year.

- **new field discoveries**
  The volumes of proved reserves of crude oil, natural gas, and/or natural gas liquids discovered in new fields during the report year.

- **new reservoir**
  A reservoir discovered during the report year.

- **nitrogen dioxide**
  A compound of nitrogen and oxygen formed by the oxidation of nitric oxide (NO) which is produced by the combustion of solid fuels.

- **nitrogen oxides (nox)**
  Compounds of nitrogen and oxygen produced by the burning of fossil fuels.

- **nitrous oxide (n2o)**
  A colorless gas, naturally occurring in the atmosphere. Nitrous oxide has a 100-year Global Warming Potential of 310.

- **no. 1 diesel fuel**
  A light distillate fuel oil that has distillation temperatures of 550 degrees Fahrenheit at the 90-percent point and meets the specifications defined in ASTM Specification D 975. It is used in high-speed diesel engines, such as those in city buses and similar vehicles.

- **no. 1 fuel oil**
  A light distillate fuel oil that has distillation temperatures of 400 degrees Fahrenheit at the 10-percent recovery point and 550 degrees Fahrenheit at the 90-percent point and meets the specifications defined in ASTM Specification D 396. It is used primarily as fuel for portable outdoor stoves and portable outdoor heaters.

- **no. 2 diesel fuel**
  A fuel that has distillation temperatures of 500 degrees Fahrenheit at the 10-percent recovery point and 640 degrees Fahrenheit at the 90-percent recovery point and meets the specifications defined in ASTM Specification D 975. It is used in high-speed diesel engines, such as those in railroad locomotives, trucks, and automobiles.

- **no. 2 distillate**
  A petroleum distillate that can be used as either a diesel fuel or a fuel oil.

- **no. 2 fuel oil (heating oil)**
  A distillate fuel oil that has distillation temperatures of 400 degrees Fahrenheit at the 10-percent recovery point and 640 degrees Fahrenheit at the 90-percent recovery point and meets the specifications defined in ASTM Specification D 396. It is used in atomizing type burners for domestic heating or for moderate capacity commercial/industrial burner units.

- **no. 2 fuel oil and no. 2 diesel sold to consumers for all other end uses**
  Those consumers who purchase fuel oil or diesel fuel for their own use including: commercial/institutional buildings (including apartment buildings), manufacturing and non-manufacturing establishments, farms (including farm houses), motor vehicles, commercial or private boats, military, governments, electric utilities, railroads, construction, logging or any other nonresidential end-use purpose.

- **no. 2 fuel oil sold to private homes for heating**
  Private household customers who purchase fuel oil for the specific purpose of heating their home, water heating, cooking, etc., excluding farm houses, farming and apartment buildings.

- **no. 4 fuel oil**
  A distillate fuel oil made by blending distillate fuel oil and residual fuel oil stocks. It conforms with ASTM Specification D 396 or Federal Specification VV-F-815C and is used extensively in industrial plants and in commercial burner installations that are not equipped with preheating facilities. It also includes No. 4 diesel fuel used for low- and medium-speed diesel engines and conforms to ASTM Specification D 975.

- **no. 5 and no. 6 fuel oil sold directly to the ultimate consumer**
  Includes ships, mines, smelters, manufacturing plants, electric utilities, drilling, railroad.

- **no. 5 and no. 6 fuel oil sold to refiners or other dealers who will resale the product**
  Includes all volumes of No. 5 and No. 6 fuel oil purchased by a trade or business with the intent of reselling the product to the ultimate consumers.

- **noaa division**
  One of the 345 weather divisions designated by the National Oceanic and Atmospheric Administration (NOAA) encompassing the 48 contiguous states. These divisions usually follow county borders to encompass counties with similar weather conditions. The NOAA

division does not follow county borders when weather conditions vary considerably within a county; such is likely to happen when the county borders the ocean or contains high mountains. A state contains an average of seven NOAA divisions; a NOAA division contains an average of nine counties.

- **noaa**
National Oceanic and Atmospheric Administration.

- **noload loss**
Power and energy lost by an electric system when not operating under demand.

- **nominal dollars**
A measure used to express nominal price.

- **nominal price**
The price paid for a product or service at the time of the transaction. Nominal prices are those that have not been adjusted to remove the effect of changes in the purchasing power of the dollar; they reflect buying power in the year in which the transaction occurred.

- **nonassociated natural gas**
Natural gas that is not in contact with significant quantities of crude oil in the reservoir.

- **nonattainment area**
Any area that does not meet the national primary or secondary ambient air quality standard established by the Environmental Protection Agency for designated pollutants, such as carbon monoxide and ozone.

- **nonbranded product**
Any refined petroleum product that is not a branded product.

- **noncoincident demand**
Sum of two or more demands on individual systems that do not occur in the same demand interval.

- **noncoincidental peak load**
The sum of two or more peak loads on individual systems that do not occur in the same time interval. Meaningful only when considering loads within a limited period of time, such as a day, week, month, a heating or cooling season, and usually for not more than 1 year.

- **nonconventional plant (uranium)**
A facility engineered and built principally for processing of uraniferous solutions that are produced during in situ leach mining, from heap leaching, or in the manufacture of other commodities, and the recovery, by chemical treatment in the plant's circuits, of uranium from the processing solutions.

- **nondedicated vehicle**
A motor vehicle capable of operating on an alternative fuel

and /or on either gasoline or diesel.

- **nonfirm power**
Power or power-producing capacity supplied or available under a commitment having limited or no assured availability.

- **nonfuel components**
Components that are not associated with a particular fuel. These include, but are not limited to, control spiders, burnable poison rod assemblies, control rod elements, thimble plugs, fission chambers, primary and secondary neutron sources, and BWR (boiling water reactor) channels.

- **nonfuel use (of energy)**
Use of energy as feedstock or raw material input.

- **nonfungible product**
A gasoline blend or blend stock that cannot be shipped via existing petroleum product distribution systems because of incompatibility problems. Gasoline/ethanol blends, for example, are contaminated by water that is typically present in petroleum product distribution systems.

- **nonhydrocarbon gases**
Typical nonhydrocarbon gases that may be present in reservoir natural gas, such as carbon dioxide, helium, hydrogen sulfide, and nitrogen.

- **nonmethane volatile organic compounds (nmvoc)**
Organic compounds, other than methane, that participate in atmospheric photochemical reactions.

- **nonoperating interest**
Any mineral lease interest (e.g., royalty, production payment, net profits interest) that does not involve the rights and obligations of operating a mineral property.

- **nonproducing reservoir**
Reservoir in which oil and/or gas proved reserves have been identified, but which did not produce during the report year to the owned or contracted interest of the reporting company regardless of the availability and/or operation of production, gathering, or transportation facilities.

- **nonrenewable fuels**
Fuels that cannot be easily made or "renewed," such as oil, natural gas, and coal.

- **nonrequirements consumer**
A wholesale consumer (unlike a full or partial requirements consumer) that purchases economic or coordination power to supplement their own or another system's energy needs.

- **nonresidential building**
A building used for some purpose other than residential. Nonresidential buildings comprise

three groups: commercial, manufacturing/industrial, and agricultural.

- **nonroad alternative fuel vehicle (nonroad afv)**

An alternative fuel vehicle designed for off-road operation and use for surface/air transportation, industrial, or commercial purposes. Nonroad AFVs include forklifts and other industrial vehicles, rail locomotives, self-propelled electric rail cars, aircraft, airport service vehicles, construction vehicles, agricultural vehicles, and marine vessels. Recreational AFVs (golf carts, snowmobiles, pleasure watercraft, etc.) are excluded from the definition.

- **nonspinning reserve**

The generating capacity not currently running but capable of being connected to the bus and load within a specified time.

- **nonutility generation**

Electric generation by end-users, or small power producers under the Public Utility Regulatory Policies Act, to supply electric power for industrial, commercial, and military operations, or sales to electric utilities.

- **nonutility power producer**

A corporation, person, agency, authority, or other legal entity or instrumentality that owns or operates facilities for electric generation and is not an electric utility. Nonutility power producers include qualifying **cogenerators**, qualifying small power producers, and other nonutility generators (including independent power producers).

- **north american electric reliability council (nerc)**

A council formed in 1968 by the electric utility industry to promote the reliability and adequacy of bulk power supply in the electric utility systems of North America. NERC consists of regional reliability councils and encompasses essentially all the power regions of the contiguous United States, Canada, and Mexico.

- **north american industry classification system (naics)**

A new classification scheme, developed by the Office of Management and Budget to replace the Standard Industrial Classification (SIC) System, that categorizes establishments according to the types of production processes they primarily use.

- **nuclear disaster 1: three mile island, united states, 1979**

  - The Three Mile Island pressurized water reactor (PWR) power station is near Harrisburg, Pennsylvania.

- Its original reactor entered service in 1974 and is still working today.
- A second reactor was destroyed in March 1979 when it had just entered service. It was the first major accident at a commercial nuclear power plant, and the worst in United States history.
- A combination of human error, mechanical failure, and poor control room design led to a loss-of-coolant accident involving partial reactor core meltdown and release of radioactivity into the

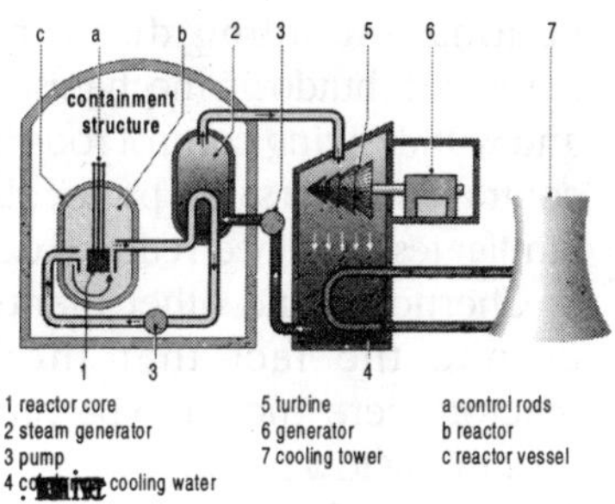

- No one was killed, and no adverse health effects caused by radiation have been found in the population around Three Mile Island.
- The average radiation dose to people within ten miles of the plant was 8 millirem, about the same as having a chest X-ray.
- The clean-up took 12 years and cost around 1 billion dollars.
- The accident caused widespread alarm and dealt a serious blow to public confidence in nuclear energy, leading to a sharp decline in nuclear construction through the 1980s and 1990s.

How a PWR works?

A pressurized water reactor (PWR) generates heat by promoting nuclear fission in the reactor core. Control rods slow the reaction by absorbing neutrons, so they are withdrawn to allow the reactor to heat up to operating temperature. Water, kept under pressure to prevent it boiling, circulates through the core and carries heat from the reactor to a heat exchanger. The heat exchanger boils separate water in the steam generator. The steam spins a turbine connected to a generator, which produces electricity.

▪ nuclear disaster 2: chernobyl, ukraine, 1986

- The accident at the Chernobyl nuclear power plant on April 26, 1986 was the most serious in the history of the nuclear industry.
- The accident was the result of a flawed reactor design, operated by inadequately trained personnel with insufficient awareness of safety.
- The resulting steam explosion and fire released about 5

percent of the radioactive reactor core into the atmosphere.

- Most radioactive material was deposited over a wide area around the plant, but a significant amount was carried downwind.
- The fallout contaminated about 58,000 square miles of Belarus, the Russian Federation and Ukraine, with a combined population of about five million people.
- Traces of radioactive fallout from Chernobyl have been measured throughout the northern hemisphere.
- 30 power plant employees and firemen were killed within days or weeks, including 28 with acute radiation syndrome.
- 700 cases of thyroid cancer, including 10 deaths, have been linked to radiation exposure resulting from the accident.
- During 1986, about 116,000 people were evacuated from around the reactor. About 220,000 people were subsequently relocated from the contaminated areas.
- A UN report in 2000 found that there is no firm scientific evidence of significant radiation-related health effects to most of the estimated 1 million people exposed, although lack of information prior to 1986 makes comparisons difficult.
- Psychosocial effects among those affected are emerging as a serious problem, similar to those arising from other major disasters such as earthquakes, floods and fires.

The accident

Prior to a routine shutdown, the crew at Chernobyl Reactor Number 4 began preparing for a test to find out how long the turbines would spin and supply electricity following a loss of power from the reactor.

They carried out a series of actions, including disabling automatic shutdown mechanisms and withdrawing control rods in contravention of safety protocols. Similar tests had been carried out at Chernobyl and other plants, despite the fact that these reactors were known to be very unstable at low power.

As the flow of coolant water was reduced, the reactor became hotter. When the operator tried to shut down the unstable reactor, there was a catastrophic power surge. This was a know design problem, resulting from the use of graphite as the moderator in combination with water as the coolant.

The fuel elements ruptured and the explosive force of super-heated steam blew the massive steel cover plate off the reactor, releasing fission products to the atmosphere.

A second explosion threw out fragments of burning fuel and graphite from the core and allowed air to rush in, causing the graphite moderator to burst into flames. The graphite burned for nine days, causing the main release of radioactivity into the environment.

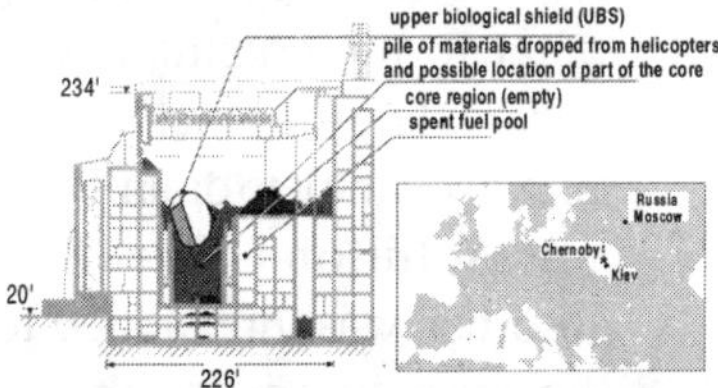

The clean-up

600 emergency workers fought to contain the fire on the night of the accident, receiving high doses of radiation.

Around 5,500 tons of boron, dolomite, sand, clay, and lead were dropped by helicopter onto the burning core in an attempt to extinguish the blaze and stop the release of radioactive particles.

About 600,000 people (civilian and military) took part in recovery operations to decontaminate the reactor block, reactor site and roads, as well as construction of the sarcophagus and of a new town for reactor personnel. These tasks were completed by 1990.

The remaining three reactors at Chernobyl resumed operation for several years, although they have now been decommissioned (the last one in December 2000).The original concrete sarcophagus sealing in the wrecked reactor is unsafe, and remedial work will cost approximately $715 million.

■ nuclear disaster 2: spread of chernobyl gases, ukraine, 1986

Material released from the Chernobyl explosion included xenon gas, iodine, caesium, and approximately 5 percent of any radioactive material remaining in the reactor after the explosion. The bulk of the material fell as dust and debris in the immediate vicinity of the reactor. Lighter materials were carried by wind over southern Russia. Some circulated over Northern Euope and the North Atlantic.

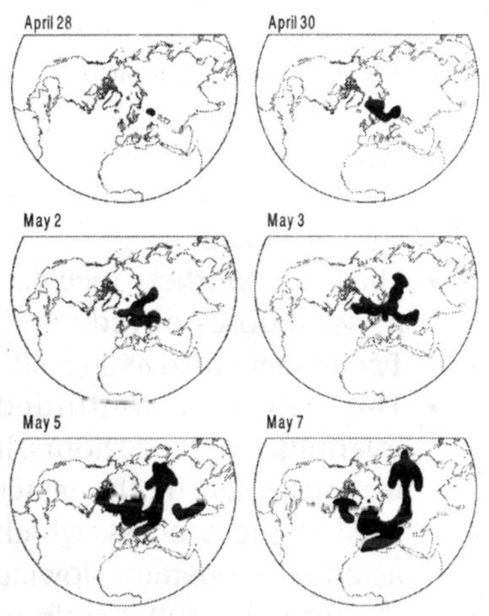

▪ nuclear disaster 3: tokaimura, Japan, 1999

- On September 30, 1999 the worst nuclear accident in the history of the Japanese nuclear power programme occurred in Tokaimura, 87 miles northwest of Tokyo.
- The plant, owned by JCO Co., was preparing a small batch of nuclear fuel for use in the experimental JOYO fast breeder reactor, using uranium enriched to 18.8 percent U-235.
- Three employees poured 35 lb of a solution containing the enriched uranium (**a**) into a precipitation tank containing nitric acid (**b**), instead of the 5.2 lb normally used.
- Bringing together so much enriched uranium triggered a limited uncontrolled nuclear chain reaction, known as a 'criticality'.
- There was no explosion, but a flash of blue light followed by alarms as the nuclear reaction emitted intense neutron and gamma radiation.
- The three workers received massive doses of radiation. Two of them died as a result.
- The criticality continued intermittently for 20 hours. It was sustained by the water used in the process, which acted as a moderator, slowing the neutrons sufficiently to keep the chain reaction going. A sleeve containing cooling water was also acting as a neutron reflector, bouncing neutrons back into the reaction.
- 161 people were evacuated from 39 houses within a 380 yard radius 5 hours after the criticality began.
- The chain reaction was finally stopped when workers crushed the pipes supplying water to the cooling sleeve surrounding the precipitation tank, allowing it to drain. Boric acid solution, a neutron absorber, was then added to the tank to ensure that it stayed subcritical.

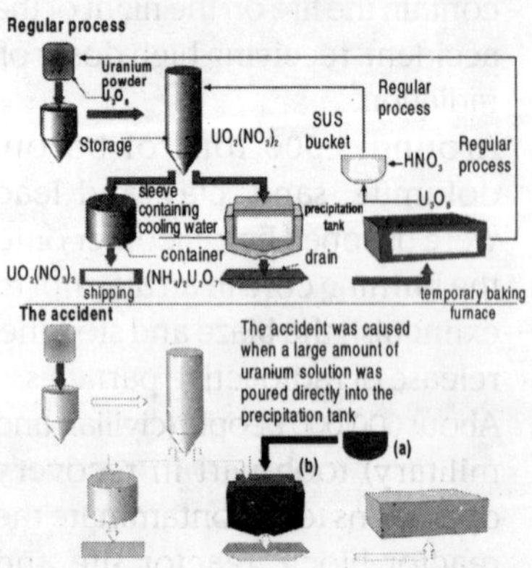

- These operations exposed a further 27 employees to some radiation.
- The International Atomic Energy Agency rated the

accident as the worst since Chernobyl, blaming it on 'human error and serious breaches of safety principles.'

- **nuclear electric power (nuclear power)**

Electricity generated by the use of the thermal energy released from the fission of nuclear fuel in a reactor.

- **nuclear energy**

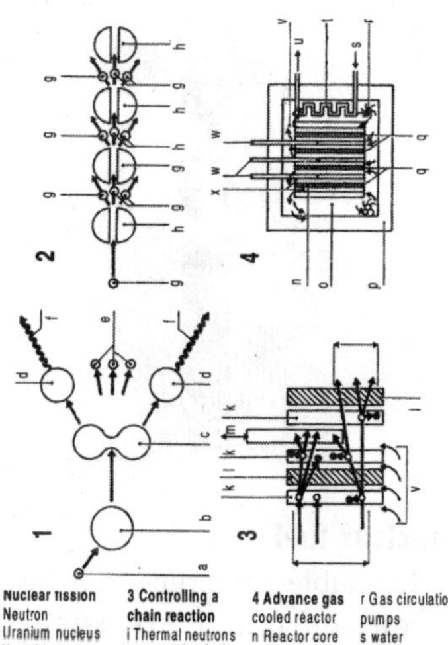

1 Nuclear fission
a Neutron
b Uranium nucleus
c Unstable nucleus
d Fission fragments, eg krypton and barium
e Neutrons
f Y -radiation
2 Chain reaction
g Neutron
h Fission of uranium nucleus

3 Controlling a chain reaction
i Thermal neutrons from another fuel rod
j Thermal neutrons to another fuel rod
k Fuel rod
l Moderator (eg graphite)
m Control rod which can be raised or lowered

4 Advance gas cooled reactor
n Reactor core up to 1500°C
o Co_2 coolant
p Concrete pressure and radiation shield
q 15000 graphite bricks locked together
r Gas circulation pumps
s water
t Steam generator
u Steam at 650°C and 40 atm pressure
v Flow of coolant
w Boron steel control rods
x Uranium fuel rods

- **nuclear energy - fission and fusion**

Another major form of energy is nuclear energy, the energy that is trapped inside each atom. One of the laws of the universe is that matter and energy can't be created nor destroyed. But they can be changed in form.

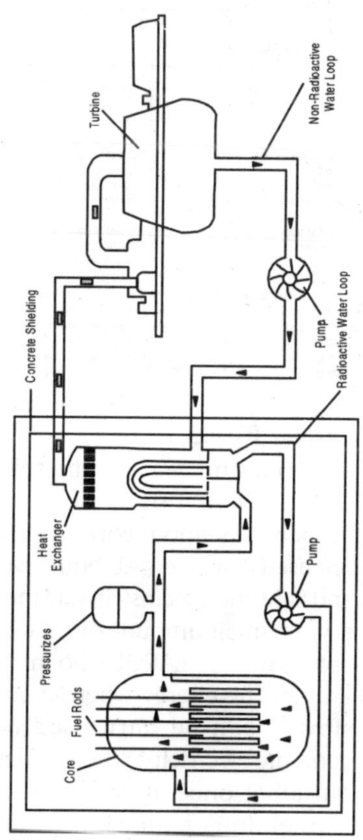

- **nuclear energy production**

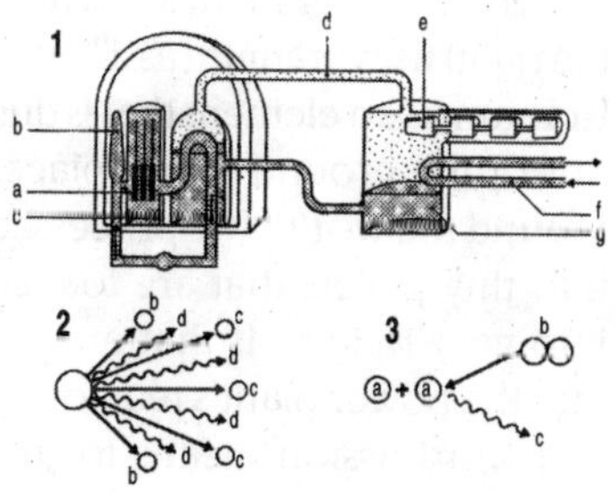

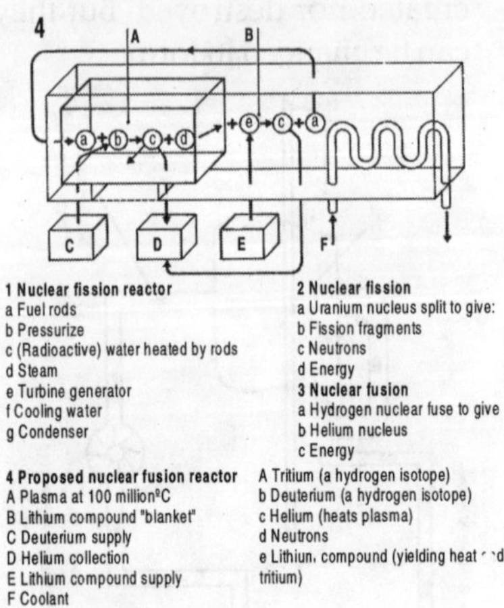

1 Nuclear fission reactor
a Fuel rods
b Pressurize
c (Radioactive) water heated by rods
d Steam
e Turbine generator
f Cooling water
g Condenser

2 Nuclear fission
a Uranium nucleus split to give:
b Fission fragments
c Neutrons
d Energy

3 Nuclear fusion
a Hydrogen nuclear fuse to give
b Helium nucleus
c Energy

4 Proposed nuclear fusion reactor
A Plasma at 100 million°C
B Lithium compound "blanket"
C Deuterium supply
D Helium collection
E Lithium compound supply
F Coolant

A Tritium (a hydrogen isotope)
b Deuterium (a hydrogen isotope)
c Helium (heats plasma)
d Neutrons
e Lithiun. compound (yielding heat [illegible]d tritium)

▪ nuclear fission

An atom's nucleus can be split apart. When this is done, a tremendous amount of energy is released. The energy is both heat and light energy. Einstein said that a very small amount of matter contains a very LARGE amount of energy. This energy, when let out slowly, can be harnessed to generate electricity. When it is let out all at once, it can make a tremendous explosion in an atomic bomb.

A nuclear power plant (like Diablo Canyon Nuclear Plant shown on the right) uses uranium as a "fuel." Uranium is an element that is dug out of the ground many places around the world. It is processed into tiny pellets that are loaded into very long rods that are put into the power plant's reactor.

The word fission means to split apart. Inside the reactor of an atomic power plant, uranium atoms are split apart in a controlled chain reaction.

In a chain reaction, particles released by the splitting of the atom go off and strike other uranium atoms splitting those. Those particles given off split still other atoms in a chain reaction. In nuclear power plants, control rods are used to keep the splitting regulated so it doesn't go too fast.

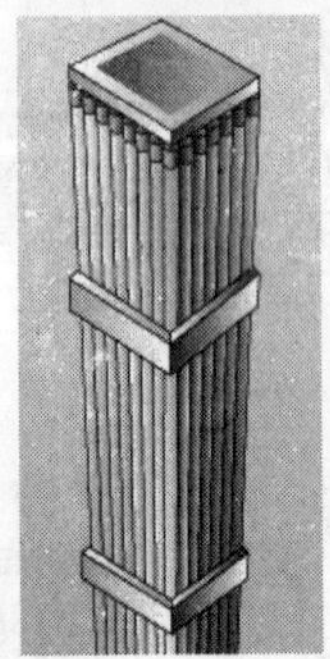

▪ nuclear fuel

Fissionable materials that have been enriched to such a composition that, when placed in a nuclear reactor, will support a self-sustaining fission chain reaction, producing heat in a controlled manner for process use.

▪ nuclear fusion

Another form of nuclear energy is called fusion. Fusion means joining smaller nuclei (the plural of nucleus) to make a larger nucleus. The sun uses nuclear fusion of hydrogen atoms into helium atoms. This gives off heat and light and other radiation.

In the picture to the right, two types of hydrogen atoms, deuterium and tritium, combine to make a helium atom and an extra particle called a neutron.

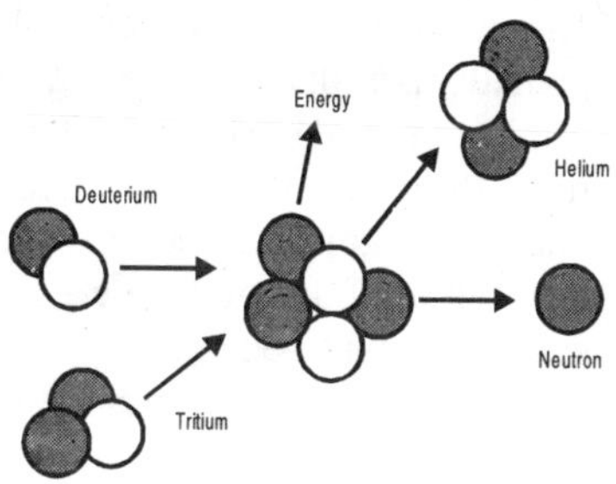

- **nuclear reactor**

An apparatus in which a nuclear fission chain reaction can be initiated, controlled, and sustained at a specific rate. A reactor includes fuel (fissionable material), moderating material to control the rate of fission, a heavy-walled pressure vessel to house reactor components, shielding to protect personnel, a system to conduct heat away from the reactor, and instrumentation for monitoring and controlling the reactor's systems.

- **nuclear waste disposal at sea**
  - During the 48-year history of the sea disposal of nuclear waste, 14 countries used more than 80 sites to dispose of some 85 PBq of radioactive waste.
  - Five countries used the sea disposal option regularly for large quantities of waste: Belgium, the former Soviet Union, Switzerland, United Kingdom, USA.

Radioactive waste dumped at sea 1946-1993
(PBq Petabecquerel [Peta=10 to the power of 15]

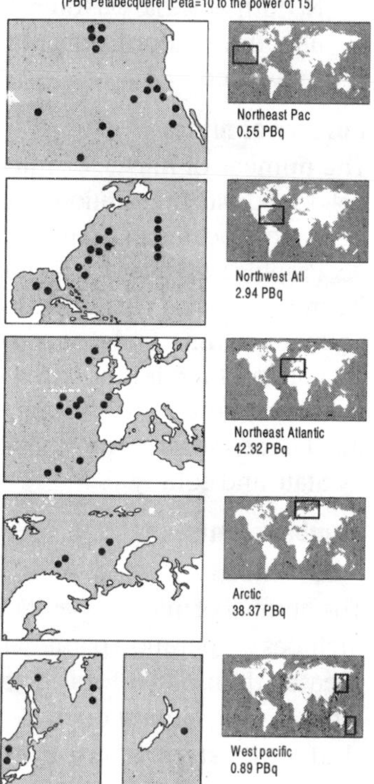

  - Some 53 percent of the radioactivity of the dumped waste is lowlevel packaged waste, most of it dumped in the Northeast Atlantic by the UK.
  - Around 43 percent is associated with the dumping of reactors with spent nuclear fuel by the former Soviet Union in the Kara Sea.

- The disposal of nuclear waste material at sea was finally banned by a resolution under the London Convention on Dumping in 1993.
- No further waste was dumped after the London Convention took place.

■ **number of mines**

The number of mines, or mines collocated with preparation plants or tipples, located in a particular geographic area (State or region). If a mine is mining coal across two counties within a State, or across two States, then it is counted as two operations. This is done so that EIA can separate production by State and county.

■ **number of mining operations**

The number of mining operations includes preparation plants with greater than 5,000 total direct labor hours. Mining operations that consist of a mine and preparation plant, or a preparation plant only, will be counted as two operations if the preparation plant processes both underground and surface coal. Excluded are silt, culm, refuse bank, slurry dam, and dredge operations except for Pennsylvania anthracite. Excludes mines producing less than 10,000 short tons of coal during the year.

■ **occupancy sensors**

These are also known as "ultrasonic switchers." When movement is detected, the lights are turned on and remain on as long as there is movement in the room.

■ **ocean energy**

The world's ocean may eventually provide us with energy to power our homes and businesses. Right now, there are very few ocean energy power plants and most are fairly small. But how can we get energy from the ocean?

There are three basic ways to tap the ocean for its energy. We can use the ocean's waves, we can use the ocean's high and low tides, or we can use temperature differences in the water.

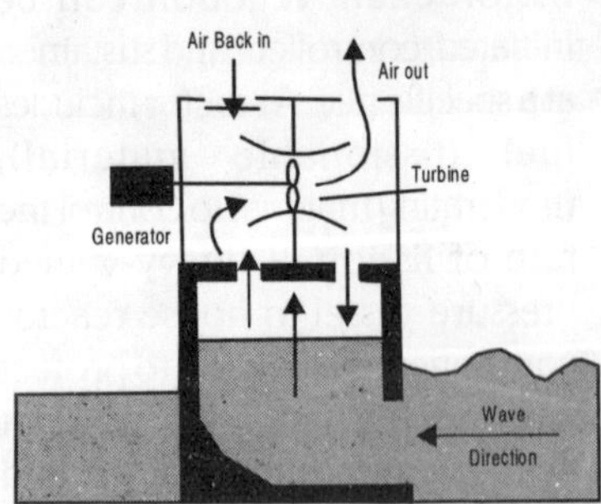

■ **ocean energy systems**

Energy conversion technologies that harness the energy in tides, waves, and thermal gradients in the oceans.

■ **ocean thermal energy conversion (OTEC)**

The process or technologies for producing energy by harnessing the temperature differences (thermal gradients) between ocean surface waters and that of ocean depths. Warm surface water is pumped through an evaporator containing a working fluid in a closed Rankine-cycle system. The vaporized fluid drives a turbine/generator.

- **octane**

A flammable liquid hydrocarbon found in petroleum. Used as a standard to measure the anti-knock properties of motor fuel.

- **octane rating**

A number used to indicate gasoline's antiknock performance in motor vehicle engines. The two recognized laboratory engine test methods for determining the antiknock rating, i.e., octane rating, of gasolines are the Research method and the Motor method. To provide a single number as guidance to the consumer, the antiknock index (R + M)/2, which is the average of the Research and Motor octane numbers, was developed.

- **oem**

Original Equipment Manufacturer.

- **off peak gas**

Gas that is to be delivered and taken on demand when demand is not at its peak.

- **off peak**

Period of relatively low system demand. These periods often occur in daily, weekly, and seasonal patterns; these off-peak periods differ for each individual electric utility.

- **off system (natural gas)**

Natural gas that is transported to the end user by the company making final delivery of the gas to the end user. The end user purchases the gas from another company, such as a producer or marketer, not from the delivering company (typically a local distribution company or a pipeline company).

- **off-highway use**

Includes petroleum products sales for use in: **1. Construction.** Construction equipment including earthmoving equipment, cranes, stationary generators, air compressors, etc. **2. Other.** Sales for off-highway uses other than construction. Sales for logging are included in this category. Volumes for off-highway use by the agriculture industry are reported under "Farm Use" (which includes sales for use in tractors, irrigation pumps, other agricultural machinery, etc.)

- **off-hours equipment reduction**

A conservation feature where there is a change in the temperature setting or reduction in the use of heating, cooling, domestic hot water heating, lighting or any other equipment either manually or automatically.

- **offshore reserves and production**

Unless otherwise dedicated, reserves and production that are in either state or Federal domains, located seaward of the coastline.

- **offshore**

That geographic area that lies seaward of the coastline. In general, the coastline is the line of ordinary low water along with that portion of the coast that is in direct contact with the open sea or the line marking the seaward limit of inland water.

- **offsite-produced energy for heat, power, and electricity generation**

This measure of energy consumption, which is equivalent to purchased energy includes energy produced off-site and consumed onsite. It excludes energy produced and consumed onsite, energy used as raw material input, and electricity losses.

- **off-system**

Any point not on, or directly interconnected with, a transportation, storage, and/or distribution system operated by a natural gas company within a state.

- **ohm**

A measure of the electrical resistance of a material equal to the resistance of a circuit in which the potential difference of 1 volt produces a current of 1 ampere.

- **ohm's law**

In a given electrical circuit, the amount of current in amperes is equal to the pressure in volts divided by the resistance, in ohms. The principle is named after the German scientist Georg Simon Ohm.

- **oil**

A mixture of hydrocarbons usually existing in the liquid state in natural underground pools or reservoirs. Gas is often found in association with oil.

- **oil and gas formation**

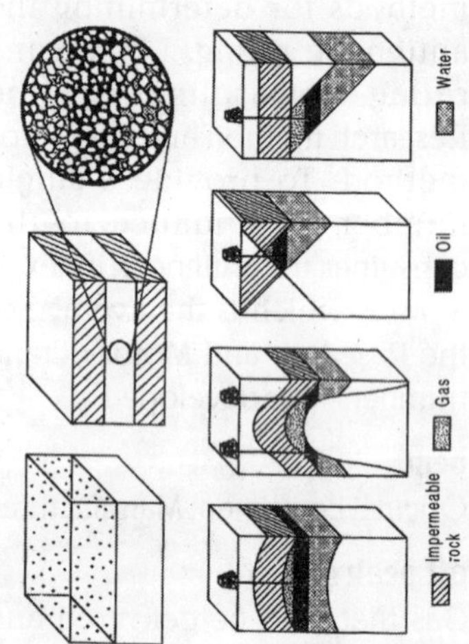

1 Formation of hydrocarbons:
1A Dead organisms sinking to the sea bed
1B Organisms in sandstone trapped by impermeable rocks
1C Bacterial action on organisms produces layer of gas, oil, and water.

2 Where oil and gas occur:
A In anticlines
B In rocks raised by salt domes
C Trapped by a Fault
D Trapped by an unconformity

■ **oil company use**

Includes sales to drilling companies, pipelines or other related oil companies not engaged in the selling of petroleum products. Includes fuel oil that was purchased or produced and used by company facilities for the operation of drilling equipment, other field or refinery operations, and space heating at petroleum refineries, pipeline companies, and oil-drilling companies. Oil used to bunker vessels is counted under vessel bunkering. Sales to other oil companies for field use are included, but sales for use as refinery charging stocks are excluded.

■ **oil reservoir**

An underground pool of liquid consisting of hydrocarbons, sulfur, oxygen, and nitrogen trapped within a geological formation and protected from evaporation by the overlying mineral strata.

■ **oil shale**

A sedimentary rock containing kerogen, a solid organic material.

■ **oil stocks**

Oil stocks include crude oil (including strategic reserves), unfinished oils, natural gas plant liquids, and refined petroleum products.

■ **oil well (casinghead) gas**

Associated and dissolved gas produced along with crude oil from oil completions.

■ **oil well**

A well completed for the production of crude oil from at least one oil zone or reservoir.

■ **old field**

A field discovered prior to the report year.

■ **old reservoir**

A reservoir discovered prior to the report year.

■ **on peak**

Periods of relatively high system demand. These periods often occur in daily, weekly, and seasonal patterns; these on-peak periods differ for each individual electric utility.

■ **one sun**

Natural solar insulation falling on an object without concentration or diffusion of the solar rays.

■ **one-time fee**

The fee assessed a nuclear utility for spent nuclear fuel (SNF) or solidified high-level radioactive waste derived from SNF, which fuel was used to generate electricity in a civilian nuclear power reactor prior to April 7, 1983, and which is assessed by applying industry-wide average dollar-per-kilogram charges to four distinct ranges of fuel burnup so that equivalent to an industry-wide average charge of 1.0 mill per kilowatt hour.

▪ **on-highway use (diesel)**
Includes sales for use in motor vehicles. Volumes used by companies in the marketing and distribution of petroleum products are also included.

▪ **onsite transportation**
The direct nonprocess end use that includes energy used in vehicles and transportation equipment that primarily consume energy within the boundaries of the establishment. Energy used in vehicles that are found primarily offsite, such as delivery trucks, is not measured by the MECS.

▪ **onsystem (natural gas)**
Natural gas that is sold (and transported) to the end user by the company making final delivery of the gas to the end user. Companies that make final delivery of natural gas are typically local distribution companies or pipeline companies.

▪ **on-system**
Any point on or directly interconnected with a transportation, storage, or distribution system operated by a natural gas company.

▪ **on-system sales**
Sales to customers where the delivery point is a point on, or directly interconnected with, a transportation, storage, and/or distribution system operated by the reporting company.

▪ **opec (organization of petroleum exporting countries)**
The acronym for the Organization of Petroleum Exporting Countries that have organized for the purpose of negotiating with oil companies on matters of oil production, prices, and future concession rights. Current members (as of the date of writing this definition) are Algeria, Indonesia, Iran, Iraq, Kuwait, Libya, Nigeria, Qatar, Saudi Arabia, the United Arab Emirates, and Venezuela.

▪ **open access**
A regulatory mandate to allow others to use a utility's transmission and distribution facilities to move bulk power from one point to another on a nondiscriminatory basis for a cost-based fee.

▪ **open market coal**
Coal is sold in the open market, i.e., coal sold to companies other than the reporting company's parent company or an operating subsidiary of the parent company.

▪ **open refrigeration unit**
Refrigeration in cabinets (units) without covers or with flexible covers made of plastic or some other material, hung in strips or curtains (fringed material, usually plastic, that push aside like a bead curtain). Flexible covers stop the

flow of warm air into the refrigerated space.

- **operable capacity**

The amount of capacity that, at the beginning of the period, is in operation; not in operation and not under active repair, but capable of being placed in operation within 30 days; or not in operation but under active repair that can be completed within 90 days. Operable capacity is the sum of the operating and idle capacity and is measured in barrels per calendar day or barrels per stream day.

- **operable generators/units**

Electric generators or generating units that are available to provide power to the grid or generating units that have been providing power to the grid but are temporarily shut down. This includes units in standby status, units out of service for an indefinite period, and new units that have their construction complete and are ready to provide test generation. A nuclear unit is operable once it receives its Full Power Operating License.

- **operable nuclear unit (foreign)**

A nuclear generating unit outside the United States that generates electricity for a grid.

- **operable nuclear unit (u.s.) A**

U.S. nuclear generating unit that has completed low-power testing and is in possession of a full-power operating license issued by the Nuclear Regulatory Commission.

- **operable refineries**

Refineries that were in one of the following three categories at the beginning of a given year: in operation; not in operation and not under active repair, but capable of being placed into operation within 30 days; or not in operation, but under active repair that could be completed within 90 days.

- **operable unit**

A unit available to provide electric power to the grid. See definition for operating unit below.

- **operable utilization rate**

Represents the use of the atmospheric crude oil distillation units. The rate is calculated by dividing the gross input to these units by the operable refining capacity of the units.

- **operated**

Exercised management responsibility for the day-to-day operations of natural gas production, gathering, treating, processing, transportation, storage, and/or distribution facilities and/or a synthetic natural gas plant.

- **operating capacity**

The component of operable capacity that is in operation at the beginning of the period.

- **operating day**
  A normal business day. Days when a company conducts business due to emergencies or other unexpected events are not included.

- **operating expenses**
  Segment expenses related both to revenue from sales to unaffiliated customers and revenue from inter segment sales or transfers, excluding loss on disposition of property, plant, and equipment; interest expenses and financial charges; foreign currency translation effects; minority interest; and income taxes.

- **operating income**
  Operating revenues less operating expenses. Excludes items of other revenue and expense, such as equity in earnings of unconsolidated affiliates, dividends, interest income and expense, income taxes, extraordinary items, and cumulative effects of accounting changes.

- **operating revenues**
  Segment revenues both from sales to unaffiliated customers (i.e., revenue from customers outside the enterprise as reported in the company's consolidated income statement) and from intersegment sales or transfers, if any, of product and services similar to those sold to unaffiliated customers, excluding equity in earnings of unconsolidated affiliates; dividend and interest income; gain on disposition of property, plant, and equipment; and foreign currency translation effects.

- **operating subsidiary**
  Company that operates a coal mining operation and is owned by another company (i.e., the parent company).

- **operating unit**
  A unit that is in operation at the beginning of the reporting period.

- **operating utilization rate**
  Represents the use of the atmospheric crude oil distillation units. The rate is calculated by dividing the gross input to these units by the operating refining capacity of the units.

- **operator, gas plant**
  The person responsible for the management and day-to-day operation of one or more natural gas processing plants as of December 31 of the report year. The operator is generally a working-interest owner or a company under contract to the working-interest owner(s). Plants shut down during the report year are also to be considered "operated" as of December 31.

- **operator, oil and/or gas well**
  The person responsible for the management and day-to-day

operation of one or more crude oil and/or natural gas wells as of December 31 of the report year. The operator is generally a working-interest owner or a company under contract to the working-interest owner(s). Wells included are those that have proved reserves of crude oil, natural gas, and/or lease condensate in the reservoirs associated with them, whether or not they are producing. Wells abandoned during the report year are also to be considered "operated" as of December 31.

- **optional delivery commitment**

A provision to allow the conditional purchase or sale of a specific quantity of material in addition to the firm quantity in the contract.

- **order**

A ruling issued by a utility commission granting or denying an application in whole or in part. The order explains the basis for the decision, noting any dispute with the factual assertions of the applicant. Also applied to a final regulation of a utility commission.

- **organic content**

The share of a substance that is of animal or plant origin.

- **organic waste**

Waste material of animal or plant origin.

- **organization for economic cooperation and development (oecd)**

An international organization helping governments tackle the economic, social and governance challenges of a globalized economy. Its membership comprises about 30 member countries. With active relationships with some 70 other countries, NGOs and civil society, it has a global reach.

- **organization of petroleum exporting countries (opec)**

Countries that have organized for the purpose of negotiating with oil companies on matters of oil production, prices, and future concession rights. Current members (as of the date of writing this definition) are Algeria, Indonesia, Iran, Iraq, Kuwait, Libya, Nigeria, Qatar, Saudi Arabia, the United Arab Emirates, and Venezuela.

- **original cost**

The initial amount of money spent to acquire an asset. It is equal to the price paid, or present value of the liability incurred, or fair value of stock issued, plus normal incidental costs necessary to put the asset into its initial use.

- **original equipment manufacturer vehicle**

A vehicle produced and marketed by an original equipment

manufacturer (OEM), including gasoline and diesel vehicles as well as alternative-fuel vehicles. A vehicle manufactured by an OEM but converted to an alternative-fuel vehicle before its initial delivery to an end-user (for example, through a contract between a conversion company and the OEM) is considered to be an OEM vehicle as long as that vehicle is still covered under the OEM's warranty.

- **original equipment manufacturer(oem)**

A company that provides the original design and materials for manufacture and engages in the assembly of vehicles. The OEM is directly responsible for manufacturing, marketing, and providing warranties for the finished product.

- **other capital costs**

Costs for items or activities not included elsewhere under capital-cost tabulations, such as for and decommissioning, dismantling, and reclamation.

- **other demand-side management (dsm) assistance programs**

A DSM program assistance that includes alternative-rate, fuel-switching, and any other DSM assistance programs that are offered to consumers to encourage their participation in DSM programs.

- **other end users**

For motor gasoline, all direct sales to end users other than those made through company outlets. For No. 2 distillate, all direct sales to end users other than residential, commercial/ institutional, industrial sales, and sales through company outlets. Included in the "other end users" category are sales to utilities and agricultural users.

- **other energy operations**

Energy operations not included under Petroleum or Coal. "Other energy" includes nuclear, oil shale, tar sands, coal liquefaction and gasification, geothermal, solar, and other forms of onconventional energy.

- **other finished**

Motor gasoline not included in the oxygenated or reformulated gasoline categories.

- **other gas**

Includes manufactured gas, coke-oven gas, blast-furnace gas, and refinery gas. Manufactured gas is obtained by distillation of coal, by the thermal decomposition of oil, or by the reaction of steam passing through a bed of heated coal or coke.

- **other generation**

Electricity originating from these sources: biomass, fuel cells, geothermal heat, solar power, waste, wind, and wood.

- **other hydrocarbons**
  Materials received by a refinery and consumed as a raw material. Includes hydrogen, coal tar derivatives, gilsonite, and natural gas received by the refinery for reforming into hydrogen. Natural gas to be used as fuel is excluded.

- **other industrial plant**
  Industrial users, not including coke plants, engaged in the mechanical or chemical transformation of materials or substances into new products (manufacturing); and companies engaged in the agriculture, mining, or construction industries.

- **other load management**
  Demand-Side Management (DSM) program other than Direct Load Control and Interruptible Load that limits or shifts peak load from on-peak to off-peak time periods. It includes technologies that primarily shift all or part of a load from one time-of-day to another and secondarily may have an impact on energy consumption. Examples include space heating and water heating storage systems, cool storage systems, and load limiting devices in energy management systems. This category also includes programs that aggressively promote time-of-use rates and other innovative rates such as real time pricing. These rates are intended to reduce consumer bills and shift hours of operation of equipment from on-peak to off-peak periods through the application of time-differentiated rates.

- **other oils equal to or greater than 401 degrees fahrenheit**
  Oils with a boiling range equal to or greater than 401 degrees Fahrenheit that are intended for use as a petrochemical feedstock.

- **other operating costs**
  Costs for other items or activities not included elsewhere in operating-cost tabulations, but required to support the calculation of a cutoff grade for ore reserves estimation.

- **other oxygenates**
  Other aliphatic alcohols and aliphatic ethers intended for motor gasoline blending (e.g., isopropyl ether (IPE) or n-propanol).

- **other power producers**
  Independent power producers that generate electricity and cogeneration plants that are not included in the other industrial, coke and commercial sectors.

- **other refiners**
  Refiners with a total refinery capacity in the United States and its possessions of less than 275,000 barrels per day as of January 1, 1982.

other service to public authorities

Electricity supplied to municipalities, divisions or agencies of state or Federal governments, under special contracts or agreements or service classifications applicable only to public authorities.

other single-unit truck

A motor vehicle consisting primarily of a single motorized device with more than two axles or more than four tires.

other supply contracts

Any contracted gas supply other than owned reserves, producer-contracted reserves, and interstate pipeline purchases that are used for acts and services for which the company has received certificate authorization from FERC. Purchases from intrastate pipelines pursuant to Section 311(b) of the NGPA of 1978 are included with other supply contracts.

other tests

Other tests include the determination of the ash softening temperature, the ash fusion temperature (the temperature at which the ash forms clinkers or slag), the free swelling index (a guide to a coal's coking characteristics), the gray king test (which determines the suitability of coal for making coke), and the hardgrove grindability index (a measure of the ease with which coal can be pulverized). in a petrographic analysis, thin sections of coal or highly polished blocks of coal are studied with a microscope to determine the physical composition, both for scientific purposes and for estimating the rank and coking potential.

other

The "other" category is defined as representing electricity consumers not elsewhere classified. This category includes public street and highway lighting service, public authority service to public authorities, railroad and railway service, and interdepartmental services.

other trucks/vans

Those trucks and vans that weigh more than 8,500 lbs GVW.

other unavailable capability

Net capability of main generating units that are unavailable for load of reasons other than full-forced outage or scheduled maintenance. Legal restrictions or other causes make these units unavailable.

outage

The period during which a generating unit, transmission line, or other facility is out of service.

outer continental shelf

Offshore Federal domain.

output

The amount of power or energy produced by a generating unit, station, or system.

oven

An appliance that is an enclosed compartment supplied with heat and used for cooking food. Toaster ovens are not considered ovens. The range stove top or burners and the oven are considered two separate appliances, although they are often purchased as one appliance.

overburden

Any material, consolidated or unconsolidated, that overlies a coal deposit.

overburden ratio

Overburden ratio refers to the amount of overburden that must be removed to excavate a given quantity of coal. It is commonly expressed in cubic yards per ton of coal, but is sometimes expressed as a ratio comparing the thickness of the overburden with the thickness of the coal bed.

overriding royalty

A royalty interest, in addition to the basic royalty, created out of the working interest; it is, therefore, limited in its duration to the life of the lease under which it is created.

owned reserves

Any reserve of natural gas that the reporting company owns as a result of oil and gas leases, fee-mineral ownership, royalty reservations, or lease or royalty reservations and assignments committed to services under certificate authorizations by FERC. Company-owned recoverable natural gas in underground storage is classified as owned reserves.

owned/rented

The relationship of a housing unit's occupants to the structure itself, not the land on which the structure is located. "Owned" means the owner or co-owner is a member of the household and the housing unit is either fully paid for or mortgaged. A household is classified "rented" even if the rent is paid by someone not living in the unit. "Rent-free" means the unit is not owned or being bought and no money is paid or contracted for rent. Such units are usually provided in exchange for services rendered or as an allowance or favor from a relative or friend not living in the unit. Unless shown separately, rent-free households are grouped with rented households.

owner occupied

Having the owner or the owner's business represented at the site. A building is considered owner occupied if an employee or representative of the owner (such as a building engineer or building manager) maintains office space in the building. Similarly, a chain store is considered owner occupied even though the actual owner may not be in the building but headquartered elsewhere. Other examples of the owner's business occupying a building include State-owned university buildings, elementary and secondary schools owned by a public school district, and a post office where the building is owned by the U.S. Postal Service.

▪ **owners equity**

Interest of the owners in the assets of the business represented by capital contributions and retained earnings.

▪ **ownership of building**

(As used in EIA's consumption surveys.) The individual, agency, or organization that owns the building. For certain EIA consumption surveys, building ownership is grouped into the following categories: Federal, State, or local government agency; a privately owned utility company; a church, synagogue, or other religious group; or any other type of individual or group.

▪ **oxidize**

To chemically transform a substance by combining it with oxygen.

▪ **oxygenated gasoline (includes gasohol)**

Finished motor gasoline, other than reformulated gasoline, having an oxygen content of 2.7 percent or higher by weight. Includes gasohol. Oxygenated gasoline excludes oxygenated fuels program reformulated gasoline (OPRG) and reformulated gasoline blendstock for oxygenate blending (RBOB).

▪ **oxygenated gasoline**

Finished motor gasoline, other than reformulated gasoline, having an oxygen content of 2.7 percent or higher by weight and required by the U.S. Environmental Protection Agency (EPA) to be sold in areas designated by EPA as carbon monoxide (CO) nonattainment areas.

▪ **oxygenates**

Substances which, when added to gasoline, increase the amount of oxygen in that gasoline blend. Ethanol, Methyl Tertiary Butyl Ether (MTBE), Ethyl Tertiary Butyl Ether (ETBE), and methanol are common oxygenates.

▪ **ozone**

A molecule made up of three atoms of oxygen. Occurs naturally in the stratosphere and provides a protective layer shielding the Earth from harmful ultraviolet radiation. In the troposphere, it is a chemical oxidant, a greenhouse gas, and a major component of photochemical smog.

▪ **ozone precursors**
Chemical compounds, such as carbon monoxide, methane, nonmethane hydrocarbons, and nitrogen oxides, which in the presence of solar radiation react with other chemical compounds to form ozone.

▪ **packaged air conditioning units**
Usually mounted on the roof or on a slab beside the building. (These are known as self-contained units, or Direct Expansion (DX). They contain air conditioning equipment as well as fans, and may or may not include heating equipment.) These are self-contained units that contain the equipment that generates cool air and the equipment that distributes the cooled air. These units commonly consume natural gas or electricity. The units are mounted on the rooftop, exposed to the elements. They typically blow cool air into the building through duct work, but other types of distribution systems may exist. The units usually serve more than one room. There are often several units on the roof of a single building. Also known as: Packaged Terminal Air Conditioners (PTAC). These packaged units are often constructed as a single unit for heating and for cooling.

▪ **packaged units**
Units built and assembled at a factory and installed as a self-contained unit to heat or cool all or portions of a building. Packaged units are in contrast to engineer-specified units built up from individual components for use in a given building. Packaged Units can apply to heating equipment, cooling equipment, or combined heating and cooling equipment. Some types of electric packaged units are also called "Direct Expansion" or DX units.

▪ **parabolic dish**
A high-temperature (above 180 degrees Fahrenheit) solar thermal concentrator, generally bowl-shaped, with two-axis tracking.

▪ **parabolic trough**
A high-temperature (above 180 degrees Fahrenheit) solar thermal concentrator with the capacity for tracking the sun using one axis of rotation.

▪ **paraffin (oil)**
A light-colored, wax-free oil obtained by pressing paraffin distillate.

▪ **paraffin (wax)**
The wax removed from paraffin distillates by chilling and pressing. When separating from solutions, it is a colorless, more or less translucent, crystalline mass, without odor and taste, slightly greasy to touch, and consisting of a mixture of solid hydrocarbons in which the paraffin series predominates.

- **paraffinic hydrocarbons**
Straight-chain hydrocarbon compounds with the general formula C_nH_{2n+2}.

- **parent**
A firm that directly or indirectly controls another entity.

- **parent company**
An affiliated company that exercises ultimate control over a business entity, either directly or indirectly, through one or more intermediaries.

- **partial requirements consumer**
A wholesale consumer with generating resources insufficient to carry all its load and whose energy seller is a long-term firm power source supplemental to the consumer's own generation or energy received from others. The terms and conditions of sale are similar to those for a full equipments consumer.

- **particulate**
A small, discrete mass of solid or liquid matter that remains individually dispersed in gas or liquid emissions. Particulates take the form of aerosol, dust, fume, mist, smoke, or spray. Each of these forms has different properties.

- **passenger-miles traveled**
The total distance traveled by all passengers. It is calculated as the product of the occupancy rate in vehicles and the vehicle miles traveled.

- **passive solar heating**
A solar heating system that uses no external mechanical power, such as pumps or blowers, to move the collected solar heat.

- **payables to municipality**
The amounts payable by the utility department to the municipality or its other departments that are subject to current settlement.

- **payment method for utilities**
The method by which fuel suppliers or utility companies are paid for all electricity, natural gas, fuel oil, kerosene, or liquefied petroleum gas used by a household. Households that pay the utility company directly are classified as "all paid by household." Households that pay directly for at least one but not all of their fuels used and that has at least one fuel charge included in the rent were classified as "some paid, some included in rent." Households for which all fuels used are included in rent were classified as "all included in rent." If the household did not fall into one of these categories, it was classified as "other." Examples of households falling into the "other" category are: (1) households for which fuel bills were paid by a social service agency or a relative, and (2) households that paid for

some of their fuels used but paid for other fuels through another arrangement.

- **peak day withdrawal**
  The maximum daily withdrawal rate (Mcf/d) experienced during the reporting period.

- **peak demand**
  The maximum load during a specified period of time.

- **peak kilowatt**
  One thousand peak watts.

- **peak load month**
  The month of greatest plant electrical generation during the winter heating season (Oct-Mar) and summer cooling season (Apr-Sept), respectively.

- **peak load plant**
  A plant usually housing old, low-efficiency steam units, gas turbines, diesels, or pumped-storage hydroelectric equipment normally used during the peak-load periods.

- **peak load**
  The maximum load during a specified period of time.

- **peak megawatt**
  One million peak watts.

- **peak watt**
  A manufacturer's unit indicating the amount of power a photovoltaic cell or module will produce at standard test conditions (normally 1,000 watts per square meter and 25 degrees Celsius).

- **peaking capacity**
  Capacity of generating equipment normally reserved for operation during the hours of highest daily, weekly, or seasonal loads. Some generating equipment may be operated at certain times as peaking capacity and at other times to serve loads on an around-the-clock basis.

- **peat**
  Peat consists of partially decomposed plant debris. It is considered an early stage in the development of coal. Peat is distinguished from lignite by the presence of free cellulose and a high moisture content (exceeding 70 percent). The heat content of air-dried peat (about 50 percent moisture) is about 9 million Btu per ton. Most U.S. peat is used as a soil conditioner. The first U.S. electric power plant fueled by peat began operation in Maine in 1990.

- **pentanes plus**
  A mixture of hydrocarbons, mostly pentanes and heavier, extracted from natural gas. Includes isopentane, natural gasoline, and plant condensate.

- **percent difference**
  The relative change in a quantity over a specified time period. It is calculated as follows: the current value has the previous value

subtracted from it; this new number is divided by the absolute value of the previous value; then this new number is multiplied by 100.

- **percent utilization**
The ratio of total production to productive capacity, times 100.

- **perfluorocarbons (pfcs)**
A group of man-made chemicals composed of one or two carbon atoms and four to six fluorine atoms, containing no chlorine. PFCs have no commercial uses and are emitted as a byproduct of aluminum smelting and semiconductor manufacturing. PFCs have very high 100-year Global Warming Potentials and are very long-lived in the atmosphere.

- **perfluoromethane**
A compound (CF_4) emitted as a byproduct of aluminum smelting.

- **permanently discharged fuel**
Spent nuclear fuel for which there are no plans for reinsertion in the reactor core.

- **permeability**
The ease with which fluid flows through a porous medium.

- **persian gulf**
The countries that surround the Persian Gulf are: Bahrain, Iran, Iraq, Kuwait, Qatar, Saudi Arabia, and the United Arab Emirates.

- **person**
An individual, a corporation, a partnership, an association, a joint-stock company, a business trust, or an unincorporated organization.

- **personal computer**
A microcomputer for producing written, programmed, or coded material; playing games; or doing calculations. Laptop and notebook computers are excluded for the purposes of EIA surveys.

- **person-year**
One whole year, or fraction thereof, worked by an employee, including contracted manpower. Expressed as a quotient (to two decimal places) of the time units worked during a year (hours, weeks, or months) divided by the like total time units in a year. For example: 80 hours worked is 0.04 (rounded) of a person-year; 8 weeks worked is 0.15 (rounded) of a person-year; 12 months worked is 1.0 person-year. Contracted manpower includes survey crews, drilling crews, consultants, and other persons who worked under contract to support a firm's ongoing operations.

- **petrochemical feedstocks**
Chemical feedstocks derived from petroleum principally for the manufacture of chemicals, synthetic rubber, and a variety of plastics.

petrochemicals

Organic and inorganic compounds and mixtures that include but are not limited to organic chemicals, cyclic intermediates, plastics and resins, synthetic fibers, elastomers, organic dyes, organic pigments, detergents, surface active agents, carbon black, and ammonia.

petroleum

A broadly defined class of liquid hydrocarbon mixtures. Included are crude oil, lease condensate, unfinished oils, refined products obtained from the processing of crude oil, and natural gas plant liquids.

petroleum administration for defense district (padd)

A geographic aggregation of the 50 States and the District of Columbia into five Districts, with PADD I further split into three subdistricts. The PADDs include the States listed below:

padd i (east coast):

- PADD IA (New England): Connecticut, Maine, Massachusetts, New Hampshire, Rhode Island, and Vermont.
- PADD IB (Central Atlantic): Delaware, District of Columbia, Maryland, New Jersey, New York, and Pennsylvania.
- PADD IC (Lower Atlantic): Florida, Georgia, North Carolina, South Carolina, Virginia, and West Virginia.
- PADD II (Midwest): Illinois, Indiana, Iowa, Kansas, Kentucky, Michigan, Minnesota, Missouri, Nebraska, North Dakota, Ohio, Oklahoma, South Dakota, Tennessee, and Wisconsin
- PADD III (Gulf Coast): Alabama, Arkansas, Louisiana, Mississippi, New Mexico, and Texas.
- PADD IV (Rocky Mountain): Colorado, Idaho, Montana, Utah, and Wyoming.
- PADD V (West Coast): Alaska, Arizona, California, Hawaii, Nevada, Oregon, and Washington.

petroleum coke, catalyst

The carbonaceous residue that is deposited on and deactivates the catalyst used in many catalytic operations (e.g., catalytic cracking). Carbon is deposited on the catalyst, thus deactivating the catalyst. The catalyst is reactivated by burning off the carbon, which is used as a fuel in the refining process. That carbon or coke is not recoverable in a concentrated form.

petroleum coke, marketable

Those grades of coke produced in delayed or fluid cokers that may be recovered as relatively pure carbon. Marketable petroleum coke may be sold as is or further purified by calcining.

- **petroleum consumption**

The sum of all refined petroleum products supplied. For each refined petroleum product, the amount supplied is calculated by adding production and imports, then subtracting changes in primary stocks (net withdrawals are a plus quantity and net additions are a minus quantity) and exports.

- **petroleum imports**

Imports of petroleum into the 50 states and the District of Columbia from foreign countries and from Puerto Rico, the Virgin Islands, and other U.S. territories and possessions. Included are imports for the Strategic Petroleum Reserve and withdrawals from bonded warehouses for onshore consumption, offshore bunker use, and military use. Excluded are receipts of foreign petroleum into bonded warehouses and into U.S. territories and U.S. Foreign Trade Zones.

- **petroleum jelly**

A semi-solid oily product produced from de-waxing lubricating oil basestocks.

- **petroleum products**

Petroleum products are obtained from the processing of crude oil (including lease condensate), natural gas, and other hydrocarbon compounds. Petroleum products include unfinished oils, liquefied petroleum gases, pentanes plus, aviation gasoline, motor gasoline, naphtha-type jet fuel, kerosene-type jet fuel, kerosene, distillate fuel oil, residual fuel oil, petrochemical feed stocks, special naphthas, lubricants, waxes, petroleum coke, asphalt, road oil, still gas, and miscellaneous products.

- **petroleum refinery**

An installation that manufactures finished petroleum products from crude oil, unfinished oils, natural gas liquids, other hydrocarbons, and alcohol.

- **petroleum stocks, primary**

For individual products, quantities that are held at refineries, in pipelines and at bulk terminals that have a capacity of 50,000 barrels or more, or that are in transit thereto. Stocks held by product retailers and resellers, as well as tertiary stocks held at the point of consumption, are excluded. Stocks of individual products held at gas processing plants are excluded from individual product stimates but are included in other oils estimates and total.

- **ph**

A measure of acidity or alkalinity. A pH of 7 represents neutrality. Acid substances have lower pH. Basic substances have higher pH.

- **photosynthesis**

The manufacture by plants of carbohydrates and oxygen from carbon dioxide and water in the presence of chlorophyll, with sunlight as the energy source. Carbon is sequestered and oxygen and water vapor are released in the process.

- **photovoltaic and solar thermal energy (as used at electric utilities)**

Energy radiated by the sun as electromagnetic waves (electromagnetic radiation) that is converted at electric utilities into electricity by means of solar (photovoltaic) cells or concentrating (focusing) collectors.

- **photovoltaic cell (pvc)**

An electronic device consisting of layers of semiconductor materials fabricated to form a junction (adjacent layers of materials with different electronic characteristics) and electrical contacts and being capable of converting incident light directly into electricity (direct current).

- **photovoltaic cell net shipments**

Represents the difference between photovoltaic cell shipments and photovoltaic cell purchases.

- **photovoltaic module**

An integrated assembly of interconnected photovoltaic cells designed to deliver a selected level of working voltage and current at its output terminals, packaged for protection against environmental degradation, and suited for incorporation in photovoltaic power systems.

- **pig iron**

Crude, high-carbon iron produced by reduction of iron ore in a blast furnace.

- **pipeline (natural gas)**

A continuous pipe conduit, complete with such equipment as valves, compressor stations, communications systems, and meters for transporting natural and/or supplemental gas from one point to another, usually from a point in or beyond the producing field or processing plant to another pipeline or to points of utilization. Also refers to a company operating such facilities.

- **pipeline (petroleum)**

Crude oil and product pipelines used to transport crude oil and petroleum products, respectively (including interstate, intrastate, and intracompany pipelines), within the 50 states and the District of Columbia.

- **pipeline freight**

Refers to freight carried through pipelines, including natural gas, crude oil, and petroleum products (excluding water). Energy is consumed by various electrical components of the pipeline, including, valves, other, appurtenances attaches to the pipe, compressor units, metering stations, regulator stations, delivery stations, holders and fabricated assemblies.

- **pipeline fuel**

Gas consumed in the operation of pipelines, primarily in compressors.

- **pipeline purchases**

Gas supply contracted from and volumes purchased from other natural gas companies as defined by the Natural Gas Act, as amended (52 Stat. 821), excluding independent producers, as defined in Paragraph 154.91(a), Chapter I, Title 18 of the Code of Federal Regulations.

- **pipeline quality natural gas**

A mixture of hydrocarbon compounds existing in the gaseous phase with sufficient energy content, generally above 900 British thermal units, and a small enough share of impurities for transport through commercial gas pipelines and sale to end-users.

- **pipeline, distribution**

A pipeline that conveys gas from a transmission pipeline to its ultimate consumer.

- **pipeline, gathering**

A pipeline that conveys gas from a production well/field to a gas processing plant or transmission pipeline for eventual delivery to end-use consumers.

- **pipeline, transmission**

A pipeline that conveys gas from a region where it is produced to a region where it is to be distributed.

- **pipelines, rate regulated**

FRS (Financial Reporting System Survey) establishes three pipeline segments: crude/liquid (raw materials); natural gas; and refined products. The pipelines included in these segments are all federally or State rate-regulated pipeline operations, which are included in the reporting company's consolidated financial statements. However, at the reporting company's option, intrastate pipeline operations may be included in the U.S. Refining/Marketing Segment if: they would comprise less than 5 percent of U.S. Refining/Marketing Segment net PP&E, revenues, and earnings in the aggregate; and if the inclusion of such pipelines in the consolidated financial statements adds less than $100 million to the net PP&E reported for the U.S. Refining/Marketing Segment.

- **pitcheblende**

Uranium oxide (U_3O_8). It is the main component of high-grade African or domestic uranium ore and also contains other oxides and sulfides, including radium, thorium, and lead components.

place in service
A vehicle is placed in service if that vehicle is new to the fleet and has not previously been in service for the fleet. These vehicles can be acquired as additional vehicles (increases the size of the company fleet), or as replacement vehicles to replace vehicles that are being retired from service (does not increase the size of the company fleet).

planetary albedo
The fraction of incident solar radiation that is reflected by the Earth-atmosphere system and returned to space, mostly by backscatter from clouds in the atmosphere.

planned generator
A proposal by a company to install electric generating equipment at an existing or planned facility or site. The proposal is based on the owner having obtained either (1) all environmental and regulatory approvals, (2) a signed contract for the electric energy, or (3) financial closure for the facility.

plant
A term commonly used either as a synonym for an industrial establishment or a generating facility or to refer to a particular process within an establishment.

plant condensate
One of the natural gas liquids, mostly pentanes and heavier hydrocarbons, recovered and separated as liquids at gas inlet separators or scrubbers in processing plants.

plant hours connected to load
The number of hours the plant is synchronized to load over a time interval usually of 1 year.

plant liquids
Those volumes of natural gas liquids recovered in natural gas processing plants.

plant or gas processing plant
A facility designated to achieve the recovery of natural gas liquids from the stream of natural gas, which may or may not have been processed through lease separators and field facilities, and to control the quality of the natural gas to be marketed.

plant products
Natural gas liquids recovered from natural gas processing plants (and in some cases from field facilities), including ethane, propane, butane, butane-propane mixtures, natural gasoline, plant condensate, and lease condensate.

plant use
The electric energy used in the operation of a plant. Included is the energy required for pumping at pump-storage plants.

plant-use electricity
The electric energy used in the operation of a plant. This energy

total is subtracted from the gross energy production of the plant.

plugged-back footage Under certain conditions, drilling operations may be continued to a greater depth than that at which a potentially productive formation is found. If production is not established at the greater depth, the well may be completed in the shallower formation. Except in special situations, the length of the well bore from the deepest depth at which the well is completed to the maximum depth drilled is defined as "plugged-back footage." Plugged-back footage is included in total footage drilled but is not reported separately.

- **plutonium (pu)**

A heavy, fissionable, radioactive, metallic element (atomic number 94) that occurs naturally in trace amounts. It can also result as a byproduct of the fission reaction in a uranium-fuel nuclear reactor and can be recovered for future use.

- **pneumatic device**

A device moved or worked by air pressure.

- **pole/tower type**

The type of transmission line supporting structure.

- **pole-mile**

A unit of measuring the simple length of an electric transmission/distribution line/feeder carrying electric conductors, without regard to the number of conductors carried.

- **polystyrene**

A polymer of styrene that is a rigid, transparent thermoplastic with good physical and electrical insulating properties, used in molded products, foams, and sheet materials.

- **polyvinyl chloride (pvc)**

A polymer of vinyl chloride. Tasteless, odorless, insoluble in most organic solvents. A member of the family vinyl resin, used in soft flexible films for food packaging and in molded rigid products, such as pipes, fibers, upholstery, and bristles.

- **pondage**

The amount of water stored behind a hydroelectric dam of relatively small storage capacity; the dam is usually used for daily or weekly control of the flow of the river.

- **pool**

In general, a reservoir. In certain situations, a pool may consist of more than one reservoir.

- **pool site**

One or more spent fuel storage pools that has a single cask loading area. Each dry cask storage area is considered a separate site.

- **population-weighted degree-days**
  Heating or cooling degree-days weighted by the population of the area in which the degree-days are recorded. To compute national population-weighted degree-days, the Nation is divided into nine Census regions comprised of from three to eight states that are assigned weights based on the ratio of the population of the region to the total population of the Nation. Degree-day readings for each region are multiplied by the corresponding population weight for each region, and these products are then summed to arrive at the national population weighted degree-day figure.

- **pore space**
  The open spaces or voides of a rock taken collectively. It is a measure of the amount of liquid or gas that may be absorbed or yielded by a particular formation.

- **portable electric heater**
  A heater that uses electricity and that can be picked up and moved.

- **portable fan**
  Box fans, oscillating fans, table or floor fans, or other fans that can be moved.

- **portable kerosene heater**
  A heater that uses kerosene and that can be picked up and moved.

- **post-mining emissions**
  Emissions of methane from coal occurring after the coal has been mined, during transport or pulverization.

- **potential consumption**
  The total amount of consumption that would have occurred had the intensity of consumption remained the same over a period of time.

- **potential peak reduction**
  The potential annual peak load reduction (measured in kilowatts) that can be deployed from Direct Load Control, Interruptible Load, Other Load Management, and Other DSM Program activities. (Please note that Energy Efficiency and Load Building are not included in Potential Peak Reduction.) It represents the load that can be reduced either by the direct control of the utility system operator or by the consumer in response to a utility request to curtail load. It reflects the installed load reduction capability, as opposed to the Actual Peak Reduction achieved by participants, during the time of annual system peak load.

- **pounds (district heat)**
  A weight quantity of steam, also used to denote a quantity of energy in the form of steam. The amount of usable energy obtained from a pound of steam depends on its temperature and

pressure at the point of consumption and on the drop in pressure after consumption.

- **power (electrical)**

An electric measurement unit of power called a voltampere is equal to the product of 1 volt and 1 ampere. This is equivalent to 1 watt for a direct current system, and a unit of apparent power is separated into real and reactive power. Real power is the work-producing part of apparent power that measures the rate of supply of energy and is denoted as kilowatts (kW). Reactive power is the portion of apparent power that does no work and is referred to as kilovars; this type of power must be supplied to most types of magnetic equipment, such as motors, and is supplied by generator or by electrostatic equipment. Voltamperes are usually divided by 1,000 and called kilovoltamperes (kVA). Energy is denoted by the product of real power and the length of time utilized; this product is expressed as kilowatt hours.

- **power ascension**

The period of time between a plant's initial fuel loading date and its date of first commercial operation (including the low-power testing period). Plants in the first operating cycle (the time from initial fuel loading to the first refueling), which lasts approximately 2 years, operate at an average capacity factor of about 40 percent.

- **power exchange**

An entity providing a competitive spot market for electric power through day- and/or hour-ahead auction of generation and demand bids.

- **power exchange generation**

Generation scheduled by the power exchange.

- **power exchange load**

Load that has been scheduled by the power exchange and is received through the use of transmission or distribution facilities owned by participating transmission owners.

- **power factor**

The ratio of real power (kilowatt) to apparent power kilovolt-ampere for any given load and time.

- **power loss**

The difference between electricity input and output as a result of an energy transfer between two points.

- **power marketers**

Business entities engaged in buying and selling electricity. Power marketers do not usually own generating or transmission facilities. Power marketers, as opposed to brokers, take ownership of the electricity and are involved in interstate trade. These entities file with the Federal Energy Regulatory Commission (FERC) for status as a power marketer.

- **power pool**

An association of two or more interconnected electric systems having an agreement to coordinate operations and planning for improved reliability and efficiencies.

- **power production plant**

All the land and land rights, structures and improvements, boiler or reactor vessel equipment, engines and engine-driven generator, turbo generator units, accessory electric equipment, and miscellaneous power plant equipment are grouped together for each individual facility.

- **power**

The rate at which energy is transferred. Electrical energy is usually measured in watts. Also used for a measurement of capacity.

- **power transfer limit**

The maximum power that can be transferred from one electric utility system to another without overloading any facility in either system.

- **powerhouse**

A structure at a hydroelectric plant site that contains the turbine and generator.

- **pp&e, additions to**

The current year's expenditures on property, plant, and equipment (PP&E). The amount is predicated upon each reporting company's accounting practice. That is, accounting practices with regard to capitalization of certain items may differ across companies, and therefore this figure in FRS (Financial Reporting System) will be a function of each reporting company's policy.

- **prediscovery costs**

All costs incurred in an extractive industry operation prior to the actual discovery of minerals in commercially recoverable quantities; normally includes prospecting, acquisition, and exploration costs and may include some development costs.

- **pregnant solution**

A solution containing dissolved extractable mineral that was leached from the ore; uranium leach solution pumped up from the underground ore zone though a production hole.

- **preliminary permit (hydroelectric power)**

A single site permit granted by the FERC (Federal Energy Regulatory Commission), which gives the recipient priority over anyone else to apply for a hydroelectric license. The preliminary permit enables the recipient to prepare a license application and conduct various studies for economic feasibility and environmental impacts. The period for a preliminary permit may extend to 3 years.

- **premium gasoline**

Gasoline having an antiknock index (R+M/2) greater than 90. Includes both leaded premium gasoline as well as unleaded premium gasoline.

- **preparation plant**

A mining facility at which coal is crushed, screened, and mechanically cleaned.

- **preproduction costs**

Costs of prospecting for, acquiring, exploring, and developing mineral reserves incurred prior to the point when production of commercially recoverable quantities of minerals commences.

- **pressurized-water reactor (pwr)**

A nuclear reactor in which heat is transferred from the core to a heat exchanger via water kept under high pressure, so that high temperatures can be maintained in the primary system without boiling the water. Steam is generated in a secondary circuit.

- **preventive maintenance program for heating and/or cooling equipment**

A HVAC conservation feature consisting of a program of routine inspection and service for the heating and/or cooling equipment. The inspection is performed on a regular basis, even if there are no apparent problems.

- **price**

The amount of money or consideration-in-kind for which a service is bought, sold, or offered for sale.

- **primary coal**

All coal milled and, when necessary, washed and sorted.

- **primary energy**

All energy consumed by end users, excluding electricity but including the energy consumed at electric utilities to generate electricity. (In estimating energy expenditures, there are no fuel-associated expenditures for hydroelectric power, geothermal energy, solar energy, or wind energy, and the quantifiable expenditures for process fuel and intermediate products are excluded.)

- **primary energy consumption expenditures**

Expenditures for energy consumed in each of the four major end-use sectors, excluding energy in the form of electricity, plus expenditures by the electric utilities sector for energy used to generate electricity. There are no fuel-associated expenditures for associated expenditures for hydroelectric power, geothermal energy, photovoltaic and solar energy, or wind energy. Also excluded are the quantifiable consumption expenditures that are an integral part of process fuel consumption.

- **primary energy consumption**
  Primary energy consumption is the amount of site consumption, plus losses that occur in the generation, transmission, and distribution of energy.

- **primary fuels**
  Fuels that can be used continuously. They can sustain the boiler sufficiently for the production of electricity.

- **primary metropolitan statistical area (pmsa)**
  A component area of a consolidated metropolitan statistical data consisting of a large urbanized county or cluster of counties (cities and towns in New England) that demonstrate strong internal economic and social links in addition to close ties with the central core of the larger area. To qualify, an area must meet specified statistical criteria that demonstrate these links and have the support of local opinion.

- **primary recovery**
  The crude oil or natural gas recovered by any method that may be employed to produce them where the fluid enters the well bore by the action of natural reservoir pressure (energy or gravity).

- **primary transportation**
  Conveyance of large shipments of petroleum raw materials and refined products usually by pipeline, barge, or ocean-going vessel. All crude oil transportation is primary, including the small amounts moved by truck. All refined product transportation by pipeline, barge, or ocean-going vessel is primary transportation.

- **prime mover**
  The engine, turbine, water wheel, or similar machine that drives an electric generator; or, for reporting purposes, a device that converts energy to electricity directly (e.g., photovoltaic solar and fuel cells).

- **prime supplier**
  A firm that produces, imports, or transports selected petroleum products across State boundaries and local marketing areas, and sells the product to local distributors, local retailers, or end users.

- **private fueling facility**
  A fueling facility which normally services only fleets and is not open to the general public.

- **privately owned electric utility**
  A class of ownership found in the electric power industry where the utility is regulated and authorized

to achieve an allowed rate of return.

- **probable (indicated) reserves, coal**

Reserves or resources for which tonnage and grade are computed partly from specific measurements, samples, or production data and partly from projection for a reasonable distance on the basis of geological evidence. The sites available are too widely or otherwise inappropriately spaced to permit the mineral bodies to be outlined completely or the grade established throughout.

- **probable energy reserves**

Estimated quantities of energy sources that, on the basis of geologic evidence that supports projections from **proved reserves**, can reasonably be expected to exist and be recoverable under existing economic and operating conditions. Site information is insufficient to establish with confidence the location, quality, and grades of the energy source. Note: This term is equivalent to "Indicated Reserves" as defined in the resource/reserve classification contained in the U.S. Geological Survey Circular 831, 1980. Measured and indicated reserves, when combined, constitute **demonstrated reserves**.

- **process cooling and refrigeration**

The direct process end use in which energy is used to lower the temperature of substances involved in the manufacturing process. Examples include freezing processed meats for later sale in the food industry and lowering the temperature of chemical feedstocks below ambient temperature for use in reactions in the chemical industries. Not included are uses such as air-conditioning for personal comfort and cafeteria refrigeration.

- **process fuel**

All energy consumed in the acquisition, processing, and transportation of energy. Quantifiable process fuel includes three categories: natural gas lease and plant operations, natural gas pipeline operations, and oil refinery operations.

- **process heating or cooling demand-side management (dsm) program**

A DSM program designed to promote increased electric energy efficiency applications in industrial process heating or cooling.

- **process heating or cooling waste heat recovery**

An energy conservation system whereby some space heating or

water heating is done by actively capturing byproduct heat that would otherwise be ejected into the environment. In non-residential buildings, sources of waste heat include refrigeration/air-conditioner compressors, manufacturing or other processes, data processing centers, lighting fixtures, ventilation exhaust air, and the occupants themselves. Not to be considered is the passive use of radiant heat from lighting, workers, motors, ovens, etc., when there are no special systems for collecting and redistributing heat.

- **processed gas**

  Natural gas that has gone through a processing plant.

- **processing gain**

  The volumetric amount by which total output is greater than input for a given period of time. This difference is due to the processing of crude oil into products which, in total, have a lower specific gravity than the crude oil processed.

- **processing loss**

  The volumetric amount by which total refinery output is less than input for a given period of time. This difference is due to the processing of crude oil into products which, in total, have a higher specific gravity than the crude oil processed.

- **processing of uranium**

  The recovery of uranium produced by nonconventional mining methods, i.e., in situ leach mining, as a byproduct of copper or phosphate mining, or heap leaching.

- **processing plant**

  A surface installation designed to separate and recover natural gas liquids from a stream of produced natural gas through the processes of condensation, absorption, adsorption, refrigeration, or other methods and to control the quality of natural gas marketed and/or returned to oil or gas reservoirs for pressure maintenance, repressuring, or cycling.

- **processing**

  Uranium-recovery operations whether at a mill, an in situ leach, byproduct plant, or other type of recovery operation.

- **producer**

  A company engaged in the production and sale of natural gas from gas or oil wells with delivery generally at a point at or near the wellhead, the field, or the tailgate of a gas processing plant. For the purpose of company classification, a company primarily engaged in the exploration for, development of, and/or production of oil and/or natural gas.

- **producer and distributor coal stocks**

  Producer and distributor coal stocks consist of coal held in stock by producers/distributors at the end of a reporting period.

- **producer contracted reserves**
  The volume of recoverable salable gas reserves committed to or controlled by the reporting pipeline company as the buyer in gas purchase contracts with the independent producer as seller, including warranty contracts, and which are used for acts and services for which the company has received certificate authorization from the Federal Energy Regulatory Commission.

- **producing property**
  A term often used in reference to a property, well, or mine that produces wasting natural resources. The term means a property that produces in paying quantities (that is, one for which proceeds from production exceed operating expenses).

- **product supplied, crude oil**
  Crude oil burned on leases and by pipelines as fuel.

- **production capacity**
  The amount of product that can be produced from processing facilities.

- **production costs**
  Costs incurred to operate and maintain wells and related equipment and facilities, including depreciation and applicable operating costs of support equipment and facilities and other costs of operating and maintaining those wells and related equipment and facilities. They become part of the cost of oil and gas produced. The following are examples of production costs (sometimes called lifting costs): costs of labor to operate the wells and related equipment and facilities; repair and maintenance costs; the costs of materials, supplies, and fuels consumed and services utilized in operating the wells and related equipment and facilities; the costs of property taxes and insurance applicable to proved properties and wells and related equipment and facilities; the costs of severance taxes. depreciation, depletion, and amortization (dd&a) of capitalized acquisition, exploration, and development costs are not production costs, but also become part of the cost of oil and gas produced along with production (lifting) costs identified above. production costs include the following subcategories of costs: well workers and maintenance; operating fluid injections and improved recovery programs; operating gas processing plants; ad valorem taxes; production or severance taxes; other, including overhead.

- **production expenses**
  Costs incurred in the production of electric power that conform to the accounting requirements of

the Operation and Maintenance Expense Accounts of the FERC Uniform System of Accounts.

- **production oil and gas**
The lifting of oil and gas to the surface and gathering, treating, field processing (as in the case of processing gas to extract liquid hydrocarbons), and field storage. The production function shall normally be regarded as terminating at the outlet valve on the lease or field production storage tank. If unusual physical or operational circumstances exist, it may be more appropriate to regard the production function as terminating at the first point at which oil, gas, or gas liquids are delivered to a main pipeline, a common carrier, a refinery, or a marine terminal.

- **production payments**
A contractual arrangement providing a mineral interest that gives the owner a right to receive a fraction of production, or of proceeds from the sale of production, until a specified quantity of minerals (or a definite sum of money, including interest) has been received.

- **production plant liquids**
The volume of liquids removed from natural gas in natural gas processing plants or cycling plants during the year.

- **production, crude oil**
The volumes of crude oil that are extracted from oil reservoirs. These volumes are determined through measurement of the volumes delivered from lease storage tanks or at the point of custody transfer, with adjustment for (1) net differences between opening and closing lease inventories and (2) basic sediment and water. Crude oil used on the lease is considered production.

- **production, lease condensate**
The volume of lease condensate produced. Lease condensate volumes include only those volumes recovered from lease or field separation facilities.

- **production, natural gas liquids**
Production of natural gas liquids is classified as follows:
    - **contract production.** natural gas liquids accruing to a company because of its ownership of liquids extraction facilities that it uses to extract liquids from gas belonging to others, thereby earning a portion of the resultant liquids.
    - **leasehold production.** natural gas liquids produced, extracted, and credited to a company's interest.
    - **contract reserves.** natural gas liquid reserves corresponding to the contract production defined above.
    - **leasehold reserves.** natural gas liquid reserves

corresponding to leasehold production defined above.

- **production, natural gas**
The volume of natural gas withdrawn from reservoirs less (1) the volume returned to such reservoirs in cycling, repressuring of oil reservoirs, and conservation operations; less (2) shrinkage resulting from the removal of lease condensate; and less (3) nonhydrocarbon gases where they occur in sufficient quantity to render the gas unmarketable. Volumes of gas withdrawn from gas storage reservoirs and native gas, which has been transferred to the storage category, are not considered production. Flared and vented gas is also considered production. (This differs from "Marketed Production" which excludes flared and vented gas.)

- **production, natural gas, dry**
The volume of natural gas withdrawn from reservoirs during the report year less (1) the volume returned to such reservoirs in cycling, repressuring of oil reservoirs, and conservation operations; less (2) shrinkage resulting from the removal of lease condensate and plant liquids; and less (3) nonhydrocarbon gases where they occur in sufficient quantity to render the gas unmarketable. Volumes of gas withdrawn from gas storage reservoirs and native gas, which has been transferred to the storage category, are not considered production. This is not the same as marketed production, because the latter also excludes vented and flared gas, but contains plant liquids.

- **production, natural gas, wet after lease separation**
The volume of natural gas withdrawn from reservoirs less (1) the volume returned to such reservoirs in cycling, repressuring of oil reservoirs, and conservation operations; less (2) shrinkage resulting from the removal of lease condensate; and less (3) nonhydrocarbon gases where they occur in sufficient quantity to render the gas unmarketable. Note: Volumes of gas withdrawn from gas storage reservoirs and native gas that has been transferred to the storage category are not considered part of production. This production concept is not the same as marketed production, which excludes vented and flared gas.

- **productive capacity**
The maximum amount of coal that a mining operation can produce or process during a period with the existing mining equipment and/or preparation plant in place, assuming that the labor and materials sufficient to utilize the plant and equipment are available, and that the market

exists for the maximum production.

- **products supplied**

Approximately represents consumption of petroleum products because it measures the disappearance of these products from primary sources, i.e., refineries, natural gas-processing plants, blending plants, pipelines, and bulk terminals. In general, product supplied of each product in any given period is computed as follows: field production, plus refinery production, plus imports, plus unaccounted-for crude oil (plus net receipts when calculated on a PAD District basis) minus stock change, minus crude oil losses, minus refinery inputs, and minus exports.

- **profit**

The income remaining after all business expenses are paid.

- **program cost**

Utility costs that reflect the total cash expenditures for the year, reported in nominal dollars, that flowed out to support DSM (demand-side management) programs. They are reported in the year they are incurred, regardless of when the actual effects occur.

- **propane (c3h8)**

A normally gaseous straight-chain hydrocarbon. It is a colorless paraffinic gas that boils at a temperature of -43.67 degrees Fahrenheit. It is extracted from natural gas or refinery gas streams. It includes all products designated in ASTM Specification D1835 and Gas Processors Association Specifications for commercial propane and HD-5 propane.

- **propane air**

A mixture of propane and air resulting in a gaseous fuel suitable for pipeline distribution.

- **propane, consumer grade**

A normally gaseous paraffinic compound (C3H8), which includes all products covered by Natural Gas Policy Act Specifications for commercial and HD-5 propane and ASTM Specification D 1835. Excludes: feedstock propanes, which are propanes not classified as consumer grade propanes, including the propane portion of any natural gas liquid mixes, i.e., butane-propane mix.

- **proportional interest in investee reserves**

The proportional interest at the end of the year in the reserves of investees that are accounted for by the equity method.

- **proposed rates**

New electric rate schedule proposed by an applicant to become effective at a future date.

- **propylene (c3h6)**

An olefinic hydrocarbon recovered from refinery

processes or petrochemical processes.

- **prospecting costs**
Direct and indirect costs incurred to identify areas of interest that may warrant detailed exploration. Such costs include those incurred for topographical, geological, and geophysical studies; rights of access to properties in order to conduct such studies, salaries, equipment, instruments, and supplies for geologists, including geophysical crews, and others conducting such studies; and overhead that can be identified with those activities.

- **prospecting**
The search for an area of probable mineralization; the search normally includes topographical, geological, and geophysical studies of relatively large areas undertaken in an attempt to locate specific areas warranting detailed exploration. Prospecting usually occurs prior to the acquisition of mineral rights.

- **proved (measured) reserves, coal**
Reserves or resources for which tonnage is computed from dimensions revealed in outcrops, trenches, workings, and drill holes and for which the grade is computed from the results of detailed sampling. The sites for inspection, sampling, and measurement are spaced so closely and the geologic character is so well defined that size, shape, and mineral content are well established. The computed tonnage and grade are judged to be accurate within limits that are stated, and no such limit is judged to be different from the computed tonnage or grade by more than 20 percent.

- **proved energy reserves**
Estimated quantities of energy sources that analysis of geologic and engineering data demonstrates with reasonable certainty are recoverable under existing economic and operating conditions. The location, quantity, and grade of the energy source are usually considered to be well established in such reserves. Note: This term is equivalent to "Measured Reserves" as defined in the resource/reserve classification contained in the U.S. Geological Survey Circular 831, 1980. Measured and indicated reserves, when combined, constitute **demonstrated reserves.**

- **proximate analysis**
determines, on an as-received basis, the moisture content, volatile matter (gases released when coal is heated), fixed carbon (solid fuel left after the volatile matter is driven off), and ash (impurities consisting of silica, iron, alumina, and other incombustible

matter). the moisture content affects the ease with which coal can be handled and burned. the amount of volatile matter and fixed carbon provides guidelines for determining the intensity of the heat produced. ash increases the weight of coal, adds to the cost of handling, and can cause problems such as clinkering and slagging in boilers and furnaces.

- **public authorities**

Electricity supplied to municipalities, divisions, or agencies of state and Federal governments, usually under special contracts or agreements that are applicable only to public authorities.

- **public authority service to public authorities**

Public authority service includes electricity supplied and services rendered to municipalities or divisions or agencies of State or Federal governments under special contracts, agreements, or service classifications applicable only to public authorities.

- **public street and highway lighting**

Electricity supplied and services rendered for the purpose of lighting streets, highways, parks, and other public places; or for traffic or other signal system service, for municipalities or other divisions or agencies of State or Federal governments.

- **public utility district**

Municipal corporations organized to provide electric service to both incorporated cities and towns and unincorporated rural areas.

- **public utility**

Enterprise providing essential public services, such as electric, gas, telephone, water, and sewer under legally established monopoly conditions.

- **public utility holding company act of 1935 (puhca)**

This act prohibits acquisition of any wholesale or retail electric business through a holding company unless that business forms part of an integrated public utility system when combined with the utility's other electric business. The legislation also restricts ownership of an electric business by non-utility corporations.

- **public utility regulatory policies act (purpa) of 1978**

One part of the National Energy Act, PURPA contains measures designed to encourage the conservation of energy, more efficient use of resources, and equitable rates. Principal among these were suggested retail rate reforms and new incentives for production of electricity by cogenerators and users of renewable resources. The Commission has primary authority for implementing several key PURPA programs.

- **publicly owned electric utility**
  A class of ownership found in the electric power industry. This group includes those utilities operated by municipalities and State and Federal power agencies.

- **pulp chips**
  Timber or residues processed into small pieces of wood of more or less uniform dimensions with minimal amounts of bark.

- **pulp wood**
  Roundwood, whole-tree chips, or wood residues.

- **pulping liquor (black liquor)**
  The alkaline spent liquor removed from the digesters in the process of chemically pulping wood. After evaporation, the liquor is burned as a fuel in a recovery furnace that permits the recovery of certain basic chemicals.

- **pumped-storage hydroelectric plant**
  A plant that usually generates electric energy during peak load periods by using water previously pumped into an elevated storage reservoir during off-peak periods when excess generating capacity is available to do so. When additional generating capacity is needed, the water can be released from the reservoir through a conduit to turbine generators located in a power plant at a lower level.

- **purchase-contract imports of uranium**
  The amount of foreign-origin uranium material that enters the United States during a survey year as reported on the "Uranium Industry Annual Survey (UIAS), Form EIA-858, as purchases of uranium ore, U3O8, natural UF6, or enriched UF6. The amount of foreign-origin uranium materials that enter the country during a survey year under other types of contracts, i.e., loans and exchanges, is excluded.

- **purchased power adjustment**
  A clause in a rate schedule that provides for adjustments to the bill when energy from another electric system is acquired and its cost varies from a specified unit base amount.

- **purchased power**
  Power purchased or available for purchase from a source outside the system.

- **purchased**
  Receipts into transportation, storage, and/or distribution facilities within a state under gas purchase contracts or agreements whether or not billing or payment occurred during the report year.

- **pure pumped-storage hydroelectric plant**

A plant that produces power only from water that has previously been pumped to an upper reservoir.

- **purpa**

The Public Utility Regulatory Policies Act of 1978, passed by the U.S. Congress. This statute requires States to implement utility conservation programs and create special markets for co-generators and small producers who meet certain standards, including the requirement that States set the prices and quantities of power the utilities must buy from such facilities.

- **pvcs that convert sunlight directly into energy**

A method for producing energy by converting sunlight using photovoltaic cells (PVCs) that are solid-state single converter devices. Although currently not in wide usage, commercial customers have a growing interest in usage and, therefore, DOE has a growing interest in the impact of PVCs on energy consumption. Economically, PVCs are competitive with other sources of electricity.

- **pyrolysis**

The thermal decomposition of biomass at high temperatures (greater than 400° F, or 200° C) in the absence of air. The end product of pyrolysis is a mixture of solids (char), liquids (oxygenated oils), and gases (methane, carbon monoxide, and carbon dioxide) with proportions determined by operating temperature, pressure, oxygen content, and other conditions.

- **quad**

Quadrillion (10 to the 15th power).

- **quadrillion**

The quantity 1,000,000,000,000,000 (10 to the 15th power).

- **qualifying facility (qf)**

A cogeneration or small power production facility that meets certain ownership, operating, and efficiency criteria established by the Federal Energy Regulatory Commission (FERC) pursuant to the Public Utility Regulatory Policies Act (PURPA).

- **quality or grade (of coal)**

An informal classification of coal relating to its suitability for use for a particular purpose. Refers to individual measurements such as heat value, fixed carbon, moisture, ash, sulfur, major, minor, and trace elements, coking properties, petrologic properties, and particular organic constituents. The individual quality elements may be aggregated in various ways to classify coal for such special purposes as metallurgical, gas, petrochemical, and blending usages.

■ **quantity wires charge**

A fee for moving electricity over the transmission and/or distribution system that is based on the quantity of electricity that is transmitted.

■ **rack sales**

Wholesale truckload sales or smaller of gasoline where title transfers at a terminal.

■ **radiant barrier**

A thin, reflective foil sheet that exhibits low radiant energy transmission and under certain conditions can block radiant heat transfer; installed in attics to reduce heat flow through a roof assembly into the living space.

■ **radiant ceiling panels**

Ceiling panels that contain electric resistance heating elements embedded within them to provide radiant heat to a room.

■ **radiant energy**

Energy that transmits away from its source in all directions.

■ **radiation**

The transfer of heat through matter or space by means of electromagnetic waves.

Radiation is energy emitted as invisible particles, waves, or rays. Radioactive atoms produce radiation as they disintegrate. Everyone is exposed to small amounts of radiation each day. Air, water, food, and sunshine are a few sources of natural background radiation. Radiation also comes from other sources, such as color televisions and medical x-rays.

People are concerned about radiation exposure because it can alter or damage the human cell structure. That is why nuclear power plants are carefully monitored and employees are trained to limit their exposure to a level that is as low as reasonably achievable.

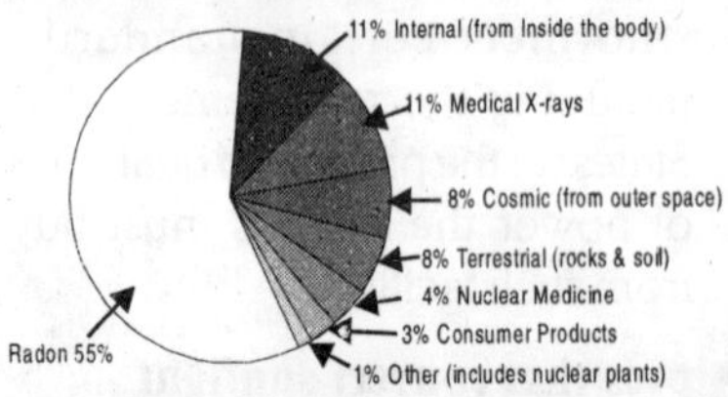

■ **radiative forcing**

A change in average net radiation at the top of the troposphere (known as the tropopause) because of a change in either incoming solar or exiting infrared radiation. A positive radiative forcing tends on average to warm the earth's surface; a negative radiative forcing on average tends to cool the earth's surface. Greenhouse gases, when emitted into the atmosphere, trap infrared energy radiated from the earth's surface and therefore tend to produce positive radiative forcing.

▪ **radiatively active gases**

Gases that absorb incoming solar radiation or outgoing infrared radiation, affecting the vertical temperature profile of the atmosphere.

▪ **radiator**

A heating unit usually exposed to view within the room or space to be heated; it transfers heat by radiation to objects within visible range and by conduction to the surrounding air, which in turn is circulated by natural convection; usually fed by steam or hot water.

▪ **radioactive waste**

Materials left over from making nuclear energy. Radioactive waste can destroy living organisms if it is not stored safely.

▪ **radioactivity**

The spontaneous emission of radiation from the nucleus of an atom. Radionuclides lose particles and energy through this process.

▪ **radioisotope**

A radioactive isotope.

▪ **radon**

A naturally occurring radioactive gas found in the United States in nearly all types of soil, rock, and water. It can migrate into most buildings. Studies have linked high concentrations of radon to lung cancer.

▪ **rail (method of transportation to consumers)**

Shipments of coal moved to consumers by rail (private or public/commercial). Includes coal hauled to or away from a railroad siding by truck.

▪ **railroad and railway electric service**

Electricity supplied to railroads and interurban and street railways, for general railroad use, including the propulsion of cars or locomotives, where such electricity is supplied under separate and distinct rate schedules.

▪ **railroad locomotive**

Self-propelled vehicle that runs on rails and is used for moving railroad cars.

▪ **railroad use**

Sales to railroads for any use, including that used for heating buildings operated by railroads.

▪ **range top**

The range burners or stove top and the oven are considered two separate appliances. Counted also with range tops are stand-alone "cook tops."

▪ **rankine cycle engine**

The Rankine cycle system uses a liquid that evaporates when heated and expands to produce work, such as turning a turbine, which when connected to a generator, produces electricity. The exhaust vapor expelled from the turbine condenses and the liquid is pumped back to the boiler to repeat the cycle. The working fluid most commonly used is water, though other liquids can also be used. Rankine cycle design is used by most commercial electric power plants. The traditional steam locomotive is also a common form of the Rankine cycle engine. The Rankine engine itself can be either a piston engine or a turbine.

- **rankine cycle**

The thermodynamic cycle that is an ideal standard for comparing performance of heat-engines, steam power plants, steam turbines, and heat pump systems that use a condensable vapor as the working fluid. Efficiency is measured as work done divided by sensible heat supplied.

- **rate base**

The value of property upon which a utility is permitted to earn a specified rate of return as established by a regulatory authority. The rate base generally represents the value of property used by the utility in providing service and may be calculated by any one or a combination of the following accounting methods: fair value, prudent investment, reproduction cost, or original cost. Depending on which method is used, the rate base includes cash, working capital, materials and supplies, deductions for accumulated provisions for depreciation, contributions in aid of construction, customer advances for construction, accumulated deferred income taxes, and accumulated deferred investment tax credits.

- **rate case**

A proceeding, usually before a regulatory commission, involving the rates to be charged for a public utility service.

- **rate features**

Special rate schedules or tariffs offered to customers by electric and/or natural gas utilities.

- **rate of return on rate base**

The ratio of net operating income earned by a utility, calculated as a percentage of its rate base.

- **rate of return**

The ratio of net operating income earned by a utility is calculated as a percentage of its rate base.

- **rate schedule (electric)**

A statement of the financial terms and conditions governing a class or classes of utility services provided to a customer. Approval of the schedule is given by the appropriate rate-making authority.

- **ratemaking authority**

A utility commission's legal authority to fix, modify, approve, or disapprove rates as determined by the powers given the commission by a State or Federal legislature.

- **rates**

The authorized charges per unit or level of consumption for a specified time period for any of the classes of utility services provided to a customer.

- **rating**

A manufacturer's guaranteed performance of a machine, transmission line, or other electrical apparatus, based on design features and test data. The rating will specify such limits as load, voltage, temperature, and frequency. The rating is generally printed on a nameplate attached to equipment and is commonly referred to as the nameplate rating or nameplate capacity.

- **ratio estimate**

The ratio of two population aggregates (totals). For example, "average miles traveled per vehicle" is the ratio of total miles driven by all vehicles, over the total number of vehicles, within any subgroup. There are two types of ratio estimates: those computed using aggregates for vehicles and those computed using aggregates for households.

- **ratoon crop**

A crop cultivated from the shoots of a perennial plant.

- **rayleigh frequency distribution**

A mathematical representation of the frequency or ratio that specific wind speeds occur within a specified time interval.

- **reactance**

A phenomenon associated with AC power characterized by the existence of a time difference between voltage and current variations.

- **reactive power**

The electrical power that oscillates between the magnetic field of an inductor and the electrical field of a capacitor. Reactive power is never converted to nonelectrical power. It is calculated as the square root of the difference between the square of the kilovolt-amperes and the square of the kilowatts and is expressed as reactive volt-amperes.

- **real dollars**

These are dollars that have been adjusted for inflation.

- **real price**

A price that has been adjusted to remove the effect of changes in the purchasing power of the dollar. Real prices, which are expressed in constant dollars, usually reflect buying power relative to a base year.

- **reasonably assured resources**

The uranium that occurs in known mineral deposits of such size, grade, and configuration that it could be recovered within the given production cost ranges, with currently proven mining and processing technology. Estimates of tonnage and grade are based on specific sample data and measurements of the deposits and on knowledge of deposit characteristics. RAR correspond to DOE's Reserves category.

- **rebate program**
  A utility company-sponsored conservation program whereby the utility company returns a portion of the purchase price cost when a more energy-efficient refrigerator, water heater, air conditioner, or other appliance is purchased.

- **reburn**
  An advanced co-firing technique using natural gas to reduce pollution from electric power plants.

- **receipts**
  1. deliveries of fuel to an electric plant 2. purchases of fuel 3. all revenues received by an exporter for the reported quantity exported

- **receivables from municipality**
  All charges by the utility department against the municipality or its other departments that are subject to current settlement.

- **received**
  Gas (and other fuels) physically transferred into the responding company's transportation, storage, and/or distribution facilities.

- **reclamation expenses**
  In the context of the coal operation statement of income, refers to all payments made by the company attributable to reclamation, including taxes.

- **reclamation**
  Process of restoring surface environment to acceptable pre-existing conditions. Includes surface contouring, equipment removal, well plugging, revegetation, etc.

- **recoverability**
  In reference to accessible coal resources, the condition of being physically, technologically, and economically minable. Recovery rates and recovery factors may be determined or estimated for coal resources without certain knowledge of their economic minability; therefore, the availability of recovery rates or factors does not predict recoverability.

- **recoverable coal**
  Coal that is, or can be, extracted from a coal bed during mining.

- **recoverable proved reserves**
  The proved reserves of natural gas as of December 31 of any

given year are the estimated quantities of natural gas which geological and engineering data demonstrates with reasonable certainty to be recoverable in the future from known natural oil and gas reservoirs under existing economic and operating conditions.

- **recoverable reserves**

The amount of coal that can be recovered (mined) from the coal deposits at active producing mines as of the end of the year.

- **recovery factor (coal)**

The percentage of total tons of coal estimated to be recoverable from a given area in relation to the total tonnage estimated to be in the demonstrated reserve base. The estimated recovery factors for the demonstrated reserve base generally are 50 percent for underground mining methods and 80 percent for surface mining methods. More precise recovery factors can be computed by determining the total coal in place and the total recoverable in any specific locale.

- **recovery percentage (coal)**

The percentage of coal that can be recovered from the coal deposits at existing mines.

- **rectifier**

A device for converting alternating current to direct current.

- **recycled feeds**

Feeds that are continuously fed back for additional processing.

- **recycling**

The process of converting materials that are no longer useful as designed or intended into a new product.

- **redox potential**

A measurement of the state of oxidation of a system.

- **redrill footage**

Occasionally, a hole is lost or junked and a second hole may be drilled from the surface in close proximity to the first. Footage drilled for the second hole is defined as "redrill footage." Under these circumstances, the first hole is reported as a dry hole (explanatory or developmental) and the total footage is reported as dry hole footage. The second hole is reported as an oil well, gas well, or dry hole according to the result. The redrill footage is included in the appropriate classification of total footage, but is not reported as a separate classification.

- **reduced use-off hours**

A conservation feature consisting of manually or automatically reducing the amount of heating or cooling produced during the hours a building is not in full use.

- **reference month**

The calendar month and year to which the reported cost, price, and volume information relates.

- **reference year**
  The calendar year to which the reported sales volume information relates.

- **refined petroleum products**
  Refined petroleum products include but are not limited to gasolines, kerosene, distillates (including No. 2 fuel oil), liquefied petroleum gas, asphalt, lubricating oils, diesel fuels, and residual fuels.

- **refiner**
  A firm or the part of a firm that refines products or blends and substantially changes products, or refines liquid hydrocarbons from oil and gas field gases, or recovers liquefied petroleum gases incident to petroleum refining and sells those products to resellers, retailers, reseller/retailers or ultimate consumers. "Refiner" includes any owner of products that contracts to have those products refined and then sells the refined products to resellers, retailers, or ultimate consumers.

- **refiner acquisition cost of crude oil**
  The cost of crude oil, including transportation and other fees paid by the refiner. The composite cost is the weighted average of domestic and imported crude oil costs. Note: The refiner acquisition cost does not include the cost of crude oil purchased for the Strategic Petroleum Reserve (SPR).

- **refinery**
  An installation that manufactures finished petroleum products from crude oil, unfinished oils, natural gas liquids, other hydrocarbons, and oxygenates.

- **refinery capacity utilization**
  Ratio of the total amount of crude oil, unfinished oils, and natural gas plant liquids run through crude oil distillation units to the operable capacity of these units.

- **refinery fuel**
  Crude oil and petroleum products consumed at the refinery for all purposes.

- **refinery gas**
  Noncondensate gas collected in petroleum refineries.

- **refinery input, crude oil**
  Total crude oil (domestic plus foreign) input to crude oil distillation units and other refinery processing units (cokers, etc.).

- **refinery input, total**
  The raw materials and intermediate materials processed at refineries to produce finished petroleum products. They include crude oil, products of natural gas processing plants, unfinished oils, other hydrocarbons and oxygenates, motor gasoline and aviation gasoline blending components and finished petroleum products.

- **refinery losses and gains**
  Processing gain and loss that takes place during the refining process

itself. Excludes losses that do not take place during the refining process, e.g., spills, fire losses, and contamination during blending, transportation, or storage.

- **refinery output**
  The total amount of petroleum products produced at a refinery. Includes petroleum consumed by the refinery.

- **refinery production**
  Petroleum products produced at a refinery or blending plant. Published production of these products equals refinery production minus refinery input. Negative production will occur when the amount of a product produced during the month is less than the amount that is reprocessed (input) or reclassified to become another product during the same month. Refinery production of unfinished oils and motor and aviation gasoline blending components appear on a net basis under refinery input.

- **refinery utilization rate**
  Represents the use of the atmospheric crude oil distillation units. The rate is calculated by dividing the gross input to these units by the operable refining capacity of the units.

- **refinery yield**
  Refinery yield (expressed as a percentage) represents the percent of finished product produced from input of crude oil and net input of unfinished oils. It is calculated by dividing the sum of crude oil and net unfinished input into the individual net production of finished products. Before calculating the yield for finished motor gasoline, the input of natural gas liquids, other hydrocarbons and oxygenates, and net input of motor gasoline blending components must be subtracted from the net production of finished aviation gasoline.

- **reflective film**
  Transparent covering for glass that helps keep out heat from the sun.

- **reflectivity**
  The ratio of the energy carried by a wave after reflection from a surface to its energy before reflection.

- **reforestation**
  Replanting of forests on lands that have recently been harvested or otherwise cleared of trees.

- **reformulated gasoline**
  Finished motor gasoline formulated for use in motor vehicles, the composition and properties of which meet the requirements of the reformulated gasoline regulations promulgated by the U.S. Environmental Protection Agency.

- **refrigeration unit**
  Lowers the temperature through a mechanical process. In a typical

refrigeration unit, electricity powers a motor that runs a pump to compress the refrigerant to maintain proper pressure. (A "refrigerant" is a substance that changes between liquid and gaseous states under desirable temperature and pressure conditions.) Heat from the compressed liquid is removed and discharged from the unit and the refrigerant then evaporates when pressure is reduced. The refrigerant picks up heat as it evaporates and it returns to the compressor to repeat the cycle. A few refrigeration units use gas (either natural gas or LPG) in an absorption process that does not use a compressor. The gas is burned to heat a chemical solution in which the refrigerant has been absorbed. Heating drives off the refrigerant which is later condensed. The condensed refrigerant evaporates by a release of pressure, and it picks up heat as it evaporates. The evaporated refrigerant is then absorbed back into the chemical solution, the heat is removed from the solution and discharged as waste heat, and the process repeats itself. By definition, refrigerators, freezers, and air-conditioning equipment all contain refrigeration units.

- **refunding**

Retirement of one security issue with proceeds received from selling another. Refunding provides for retiring maturing debt by taking advantage of favorable money market conditions.

- **refuse bank**

A repository for waste material generated by the coal cleaning process.

- **refuse mine**

A surface mine where coal is recovered from previously mined coal. It may also be known as a silt bank, culm bank, refuse bank, slurry dam, or dredge operation.

- **refuse-derived fuel (rdf)**

A fuel produced by shredding municipal solid waste (MSW). Noncombustible materials such as glass and metals are generally removed prior to making RDF. The residual material is sold as-is or compressed into pellets, bricks, or logs. RDF processing facilities are typically located near a source of MSW, while the RDF combustion facility can be located elsewhere.

- **regional transmission group**

A utility industry concept that the Federal Energy Regulatory Commission (FERC) embraced for the certification of voluntary groups that would be responsible for transmission planning and use on a regional basis.

- **regular grade gasoline**

A grade of unleaded gasoline with a lower octane rating (approx. 82) than other grades. Octane boosters are added to gasoline to control engine pre-ignition or "knocking" by slowing combustion rates.

- **regulated entity**

For the purpose of EIA's data collection efforts, entities that either provide electricity within a designated franchised service area and/or file forms listed in the Code of Federal Regulations, Title 18, part 141 are considered regulated entities. This includes investor-owned electric utilities that are subject to rate regulation, municipal utilities, federal and state power authorities, and rural electric cooperatives. Facilities that qualify as cogenerators or small power producers under the Public Utility Regulatory Power Act (PURPA) are not considered regulated entities.

- **regulated stream flow**

The rate of flow past a given point during a specified period that is controlled by reservoir water release operation.

- **regulation**

The governmental function of controlling or directing economic entities through the process of rulemaking and adjudication.

- **regulation, procedures and practices**

A utility commission carries out its regulatory functions through rulemaking and adjudication. Under rulemaking, the utility commission may propose a general rule of regulation change. By law, it must issue a notice of the proposed rule and a request for comments is also made; the Federal Energy Regulatory Commission publishes this in the Federal Register. The final decision must be published. A utility commission may also work on a case-by-case basis from submissions from regulated companies or others. Objections to a proposal may come from the commission or intervenors, in which case the proposal must be presented to a hearing presided over by an administrative law judge. The judge's decision may be adopted, modified, or reversed by the utility commissioners, in which case those involved can petition for a rehearing and may appeal a decision through the courts system to the U.S. Supreme Court.

- **reheating coils**

A part of some air conditioning systems. Electric coils in air ducts used primarily to raise the temperature of circulated air after it was over-cooled to remove moisture. Some buildings have reheating coils as their sole heating source.

- **reinjected**

The forcing of gas under pressure into an oil reservoir in an attempt to increase recovery.

■ **reinserted fuel**

Irradiated fuel that is discharged in one cycle and inserted in the same reactor during a subsequent refueling. In a few cases, fuel discharged from one reactor has been used to fuel a different reactor.

■ **reinsertion**

The process of returning nuclear fuel that has been irradiated and then removed from a reactor back into a reactor for further irradiation.

Reinserted assemblies are assemblies that have been irradiated in a cycle, were not in the core in the prior cycle (cycle N), and which are in the core in the current cycle (cycle N+1).

■ **reliability (electric system)**

A measure of the ability of the system to continue operation while some lines or generators are out of service. Reliability deals with the performance of the system under stress.

■ **relocation of tailings**

Relocation of tailings is sometimes necessary if the pile poses a threat to inhabitants or the environment, for example, through being situated too close to populated areas, on top of aquifers or other sources of water, or in unstable areas such as flood plains or faults near earthquake zones.

■ **renewable energy hydroelectric**

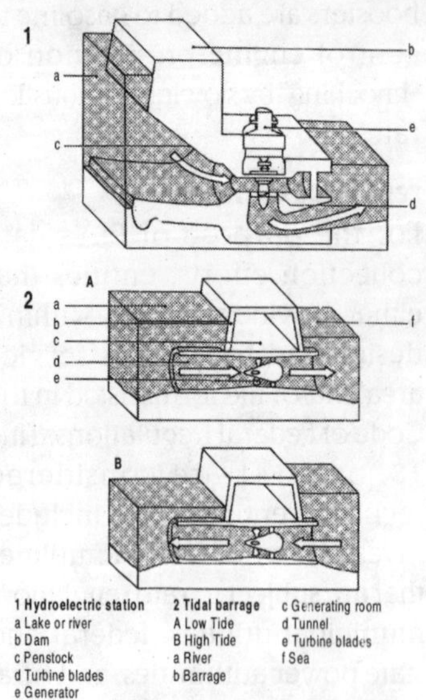

■ **renewable energy resources**

Energy resources that are naturally replenishing but flow-limited. They are virtually inexhaustible in duration but limited in the amount of energy that is available per unit of time. Renewable energy resources include: biomass, hydro, geothermal, solar, wind, ocean thermal, wave action, and tidal action.

■ **renewable energy: geothermal**

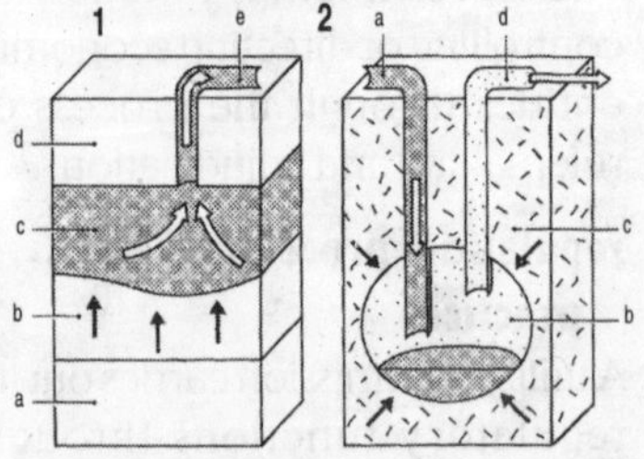

area heated, and without substantially reducing the temperature maintained in the heated area.

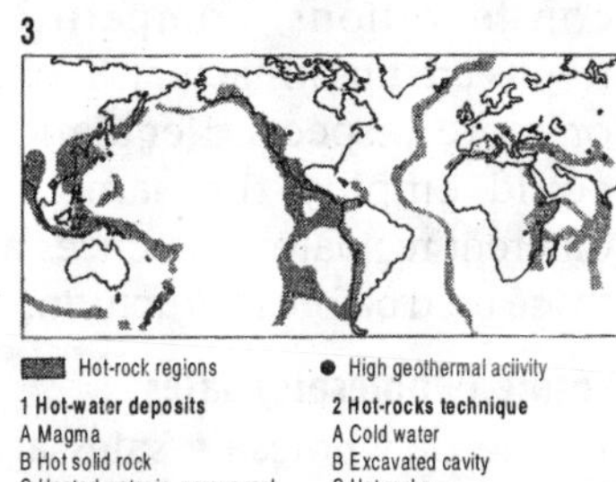

Hot-rock regions ● High geothermal activity
1 Hot-water deposits
A Magma
B Hot solid rock
C Heated water in porous rock
D Impermeable rock
E Pumped up hot water
2 Hot-rocks technique
A Cold water
B Excavated cavity
C Hot rocks
D High-pressure steam
3The World's hot rocks

■ **renewable energy: solar and wind**

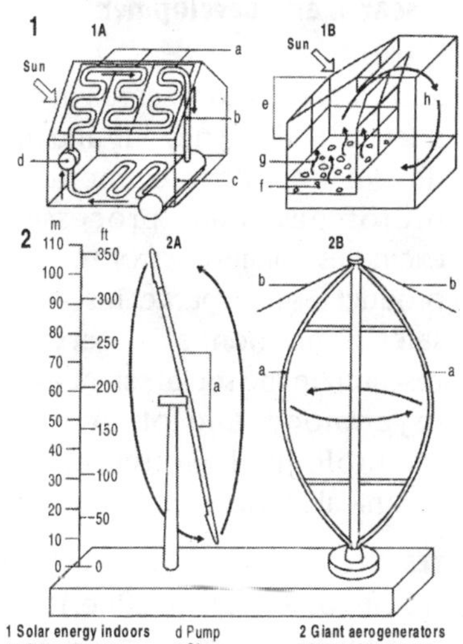

1 Solar energy indoors
1A Solar collectors
1B Passive heating system
a Solar collector panels
b Circulating water
c Storage tank
d Pump
e Glass
f Brick or stone
g Radiant and conducted heat
h Convected heat
2 Giant aerogenerators
2A Mod-5B (horizontal axis rotor)
2B EOLE (vertical axis rotor)
a Rotor
b Guy cables

■ **replacement energy source for primary heating**

For the CBECS, the heating energy source to which the building could switch within one week without major modifications to the main heating equipment, without substantially reducing the

■ **replacement vehicle**

A vehicle which is acquired in order to take the place of a vehicle which is being retired from service. These acquisitions do not increase the size of the company fleet.

■ **report state**

The State, including adjacent offshore continental shelf areas in the Federal domain, in which a company operated natural gas gathering, transportation, storage, and/or distribution facilities or a synthetic natural gas plant covered by the individual report.

■ **report week**

A calendar week beginning at 12:01 a.m. on Sunday and ending at midnight on Saturday.

■ **report year (calendar)**

The 12-month period, January 1 through December 31.

■ **report year (fiscal)**

A 12-month period for which an organization plans the use of its funds. The fiscal year is designated by the calendar year in which it ends.

■ **reporting**

The average number of Btu per cubic foot of gas at 60 degrees Fahrenheit and 14.73 psia delivered directly to consumers. Where billing is on a thermal basis, the heat content values used for billing purposes are to be used to determine the annual average heat content.

- **repowering**
  Refurbishment of a plant by replacement of the combustion technology with a new combustion technology, usually resulting in better performance and greater capacity.
- **repressuring**
  The injection of gas into oil or gas formations to effect greater ultimate recovery.
- **reprocessing**
  Synonymous with chemical separations.
- **requirements power**
  The firm service needs required by designated load plus losses from the points of supply.
- **reregulation**
  The design and implementation of regulatory practices to be applied to the remaining regulated entities after restructuring of the vertically-integrated electric utility. The remaining regulated entities would be those that continue to exhibit characteristics of a natural monopoly, where imperfections in the market prevent the realization of more competitive results, and where, in light of other policy considerations, competitive results are unsatisfactory in one or more respects. Regulation could employ the same or different regulatory practices as those used before restructuring.
- **resale (wholesale) sales**
  Resale or wholesale sales are electricity sold (except under exchange agreements) to other electric utilities or to public authorities for resale distribution.
- **research and development (r&d)**
  Basic and applied research in the sciences and engineering and the design and development of prototypes and processes, excluding quality control, routine product testing, market research, sales promotion, sales service, research in the social sciences or psychology, and other non-technological activities or technical services.
- **reseller**
  A **firm** (other than a refiner) that is engaged in a trade or business that buys refined petroleum products and then sells them to a purchaser who is not the ultimate consumer of those refined products.
- **reserve additions**
  The estimated original, recoverable, salable, and new proved reserves credited to new fields, new reservoirs, new gas

purchase contracts, amendments to old gas purchase contracts, or purchase of gas reserves in-place that occurred during the year and had not been previously reported. Reserve additions refer to domestic in-the-ground natural gas reserve additions and do not refer to interstate pipeline purchase agreements; contracts with foreign suppliers; coal gas, SNG, or LNG purchase arrangements.

- **reserve cost categories of $15, $30, $50, and $100 per pound u3o8**

Classification of uranium reserves estimated by using break-even cutoff grades that are calculated based on forward-operating costs of less than $15, $30, $50, and $100 per pound U_3O_8.

- **reserve generating capacity**

Amount of generating capacity available to meet peak or abnormally high demands for power and to generate power during scheduled or unscheduled outages.

- **reserve margin (operating)**

The amount of unused available capability of an electric power system (at peak load for a utility system) as a percentage of total capability.

- **reserve revisions**

Changes to prior year-end proved reserves estimates, either positive or negative, resulting from new information other than an increase in proved acreage (extension). Revisions include increases of proved reserves associated with the installation of improved recovery techniques or equipment. They also include correction of prior year arithmetical or clerical errors and adjustments to prior year-end production volumes to the extent that these alter reserves estimates.

- **reserve**

That portion of the demonstrated reserve base that is estimated to be recoverable at the time of determination. The reserve is derived by applying a recovery factor to that component of the identified coal resource designated as the demonstrated reserve base.

- **reserves changes**

Positive and negative revisions, extensions, new reservoir discoveries in old fields, and new field discoveries that occurred during the report year.

- **reserves, coal**

Quantities of unextracted coal that comprise the demonstrated base for future production, including both proved and probable reserves.

- **reservoir**

A porous and permeable underground formation

containing an individual and separate natural accumulation of producible hydrocarbons (crude oil and/or natural gas) which is confined by impermeable rock or water barriers and is characterized by a single natural pressure system.

- **reservoir capacity**
  The present total developed capacity (base and working) of the storage reservoir, excluding contemplated future development.

- **reservoir repressuring**
  The injection of a pressurized fluid (such as air, gas, or water) into oil and gas reservoir formations to effect greater ultimate recovery.

- **residential building**
  A structure used primarily as a dwelling for one or more households.

- **residential consumers**
  Consumers using gas for heating, air conditioning, cooking, water heating, and other residential uses in single and multi-family dwellings and apartments and mobile homes.

- **residential energy consumption survey (recs)**
  A national multistage probability sample survey conducted by the Energy End Use Division of the Energy Information Administration. The RECS provides baseline information on how households in the United States use energy. The Residential Transportation Energy Consumption Survey (RTECS) sample is a subset of the RECS. Household demographic characteristics reported in the RTECS publication are collected during the RECS personal interview.

- **residential heating oil price**
  The price charged for home delivery of No. 2 heating oil, exclusive of any discounts such as those for prompt cash payment. Prices do not include taxes paid by the consumer.

- **residential propane price**
  The "bulk keep full" price for home delivery of consumer-grade propane intended for use in space heating, cooking, or hot water heaters in residences.

- **residential sector**
  An energy-consuming sector that consists of living quarters for private households. common uses of energy associated with this sector include space heating, water heating, air conditioning, lighting, refrigeration, cooking, and running a variety of other appliances. the residential sector excludes institutional living quarters.

- **residential type central air conditioner**
  There are four basic parts to a residential central air-conditioning system: (1) a condensing unit, (2) a cooling coil,

(3) ductwork, and (4) a control mechanism such as a thermostat. There are two basic configurations of residential central systems: (1) a "split system" where the condensing unit is located outside and the other components are inside, and (2) a packaged-terminal air-encased in one unit and is usually found in a "utility closet."

■ residential vehicles

Motorized vehicles used by U.S. households for personal transportation. Excluded are motorcycles, mopeds, large trucks, and buses. Included are automobiles, station wagons, passenger vans, cargo vans, motor homes, pickup trucks, and jeeps or similar vehicles. In order to be included, vehicles must be (1) owned by members of the household, or (2) company cars not owned by household members but regularly available to household members for their personal use and ordinarily kept at home, or (3) rented or leased for 1 month or more.

■ residential/commercial (consumer category)

Housing units, wholesale or retail businesses (except coal wholesale dealers); health institutions (hospitals, social and educational institutions (schools and universities); and Federal, state, and local governments (military installations, prisons, office buildings, etc.). Excludes shipments to Federal power projects, such as TVA, and rural electrification cooperatives, power districts, and state power projects.

■ residual fuel oil

A general classification for the heavier oils, known as No. 5 and No. 6 fuel oils, that remain after the distillate fuel oils and lighter hydrocarbons are distilled away in refinery operations. It conforms to ASTM Specifications D 396 and D 975 and Federal Specification VV-F-815C. No. 5, a residual fuel oil of medium viscosity, is also known as Navy Special and is defined in Military Specification MIL-F-859E, including Amendment 2 (NATO Symbol F-770). It is used in steam-powered vessels in government service and inshore power plants. No. 6 fuel oil includes Bunker C fuel oil and is used for the production of electric power, space heating, vessel bunkering, and various industrial purposes.

■ residue gas

Natural gas from which natural gas processing plant liquid products and, in some cases, non-hydrocarbon components have been extracted.

■ residuum

Residue from crude oil after distilling off all but the heaviest components, with a boiling range greater than 1,000 degrees Fahrenheit.

- **respondent**
A company or individual who completes and returns a report or survey form.

- **restoration time**
The time when the major portion of the interrupted load has been restored and the emergency is considered to be ended. However, some of the loads interrupted may not have been restored due to local problems.

- **restricted-universe census**
This is the complete enumeration of data from a specifically defined subset of entities including, for example, those that exceed a given level of sales or generator nameplate capacity.

- **restructuring**
The process of replacing a monopoly system of electric utilities with competing sellers, allowing individual retail customers to choose their electricity supplier but still receive delivery over the power lines of the local utility. It includes the reconfiguration of the vertically-integrated electric utility.

- **retail motor gasoline prices**
Motor gasoline prices calculated each month by the Bureau of Labor Statistics (BLS) in conjunction with the construction of the Consumer Price Index.

- **retail wheeling**
The process of moving electric power from a point of generation across third-party-owned transmission and distribution systems to a retail customer.

- **retailer**
A firm (other than a refiner, reseller, or reseller/retailer) that carries on the trade or business of purchasing refined petroleum products and reselling them to ultimate consumers without substantially changing their form.

- **retained earnings**
The balance, either debit or credit, of appropriated or un-appropriated retained earnings of the utility department arising from earnings.

- **retire from service**
A vehicle is retired from service if that vehicle is placed out of service and there are no future plans to return that vehicle to service.

- **retired hydropower plant sites**
The site of a plant that formerly produced electrical or mechanical power but is now out of service. Includes plants that have been abandoned, damaged by flood or fire, inundated by new reservoirs, or dismantled.

- **return on common equity**
The net income less preferred stock dividends, divided by the average common stock equity.

return on common stock equity
An equity's earnings available for common stockholders calculated as a percentage of its common equity capital.

revenue - (electricity)
The total amount of money received by an entity from sales of its products and/or services; gains from the sales or exchanges of assets, interest, and dividends earned on investments; and other increases in the owner's equity, except those arising from capital adjustments.

revenue requirement
The total revenue that the utility is authorized an opportunity to recover, which includes operating expenses and a reasonable return on rate base.

reversible turbine
A hydraulic turbine, normally installed in a pumped-storage plant, which can be used alternatively as a pump or as an engine, turbine, water wheel, or other apparatus that drives an electrical generator.

revisions and additions (gross change in reserves)
The difference (plus or minus) between the year-end reserves plus production for a given year and the year-end reserves for the previous year.

ribbon silicon
Crystalline silicon that is used in photovoltaic cells. Ribbon silicon is fabricated by a variety of solidification (crystallization) methods that withdraw thin silicon sheets from pools of relatively pure molten silicon.

right-of-way
The land and legal right to use and service the land along which a transmission line is located. Transmission line right-of-way is usually acquired in widths that vary with the kilovolt (kV) size of the line.

rip rap
Cobblestone or coarsely broken rock used for protection against erosion of embankment or gully.

river (method of transportation to consumers - coal)
Shipments of coal moved to consumers via river by barge. Shipments to Great Lakes coal loading docks or Tidewater pier or coastal points are not included.

road oil
Any heavy petroleum oil, including residual asphaltic oil used as a dust pallative and surface treatment on roads and highways. It is generally produced in six grades, from 0, the most liquid, to 5, the most viscous.

rodlet or gad basket
An open garbage and debris (GAD) basket that may have contain pieces of fuel rods, disassembled fuel rods, and other fuel and non-fuel components.

roll front
A type of uranium deposition localized as a roll or interface separating an oxidized interior from a reduced exterior. The reduced side of this interface is significantly enriched in uranium.

roof (coal)
The rock immediately above a coal seam. The roof is commonly a shale, often carbonaceous and softer than rocks higher up in the roof strata.

roof insulation
Insulating materials placed underneath the roof or on the roof (building).

roof or ceiling insulation
A building shell conservation feature consisting of insulation placed in the roof (below the waterproofing layer) or in the ceiling of the top floor in the building.

roof or ceiling insulation, insulation in exterior walls
Any material that when placed between the interior surface of the building and the exterior surface of the building, reduces the rate of heat loss to the environment or heat gain from the environment. Roof or ceiling insulation refers to insulation placed in the roof or ceiling of the top occupied floor in the building. Wall insulation refers to insulation placed between the exterior and interior walls of the building.

roof pond
A solar energy collection device consisting of containers of water located on a roof that absorb solar energy during the day so that the heat can be used at night or that cools a building by evaporation at night.

room air conditioner
Air-conditioning units that typically fit into the window or wall and are designed to cool only one room.

room heater burning gas, oil, and kerosene
Any of the following heating equipment: circulating heaters, convectors, radiant gas heaters, space heaters, or other nonportable room heaters that may or may not be connected to a flue, vent, or chimney.

room-and-pillar mining
The most common method of underground mining in which the mine roof is supported mainly by coal pillars left at regular intervals. Rooms are places where the coal is mined; pillars are areas of coal left between the rooms. Room-

and-pillar mining is done either by conventional or continuous mining.

- **rotary rig**
  A machine used for drilling wells that employs a rotating tube attached to a bit for boring holes through rock.

- **round test mesh**
  A sieving screen with round holes, the dimensions of which are of specific sizes to allow certain sizes of coal to pass through while retaining other sizes.

- **roundwood**
  Wood cut specifically for use as a fuel.

- **royalty**
  A contractual arrangement providing a mineral interest that gives the owner a right to a fractional share of production or proceeds there from, that does not contain rights and obligations of operating a mineral property, and that is normally free and clear of exploration, developmental and operating costs, except production taxes.

- **royalty cost**
  A share of the profit or product reserved by the grantor of a mining lease, such as a royalty paid to a lessee.

- **royalty interest (including overriding royalty)**
  These interests entitle their owner(s) to a share of the mineral production from a property or to a share of the proceeds there from. They do not contain the rights and obligations of operating the property and normally do not bear any of the costs of exploration, development, and operation of the property.

- **royalty interest**
  An interest in a mineral property provided through a royalty contract.

- **rulemaking (regulations)**
  The authority delegated to administrative agencies by Congress or State legislative bodies to make rules that have the force of law. Frequently, statutory laws that express broad terms of a policy are implemented more specifically by administrative rules, regulations, and practices.

- **run off**
  That portion of the precipitation that flows over the land surface and ultimately reaches streams to complete the water cycle. Melting snow is an important source of this water as well as all amounts of surface water that move to streams or rivers through any given area of a drainage basin.

- **running and quick-start 4capability**
  The net capability of generating units that carry load or have quick-start capability. In general, quick-start capability refers to

generating units that can be available for load within a 30-minute period.

- **run-of-mine coal**
Coal as it comes from the mine prior to screening or any other treatment.

- **run-of-river hydroelectric plant**
A low-head plant using the flow of a stream as it occurs and having little or no reservoir capacity for storage.

- **rural electrification administration (rea)**
A lending agency of the U. S. Department of Agriculture, the REA makes self-liquidating loans to qualified borrowers to finance electric and telephone service to rural areas. The REA finances the construction and operation of generating plants, electric transmission and distribution lines, or systems for the furnishing of initial and continued adequate electric services to persons in rural areas not receiving central station service.

- **r-value**
A measure of a material's resistance to heat flow in units of Fahrenheit degrees x hours x square feet per Btu. The higher the R-value of a material, the greater its insulating capability. The R-value of some insulating materials is 3.7 per inch for fiberglass and cellulose, 2.5 per inch for vermiculite, and more than 4 per inch for foam. All building materials have some R-value. For example, a 4-inch brick has an R-value of 0.8, and half-inch plywood has an R-value of 0.6.

- **salable coal**
The shippable product of a coal mine or preparation plant. Depending on customer specifications, salable coal may be run-of-mine, crushed-and-screened (sized) coal, or the clean coal yield from a preparation plant.

- **salable natural gas**
Natural gas marketed under controlled quality conditions.

- **sales for resale**
A type of wholesale sales covering energy supplied to other electric utilities, cooperatives, municipalities, and Federal and state electric agencies for resale to ultimate consumers.

- **sales to end users**
Sales made directly to the consumer of the product. Includes bulk consumers, such as agriculture, industry, and utilities, as well as residential and commercial consumers.

- **sales type**
Sales categories of sales to end-users and sales for resale.

- **sales volume (coal)**
The reported output from Federal and/or Indian lands, the basis of

royalties. It is approximately equivalent to production, which includes coal sold, and coal added to stockpiles.

- **salt dome**

A domical arch (anticline) of sedimentary rock beneath the earth's surface in which the layers bend downward in opposite directions from the crest and that has a mass of rock salt as its core.

- **salt gradient solar ponds**

These consist of three main layers. The top layer is near ambient and has low salt content. The bottom layer is hot, typically 160° F to 212° F (71° C to 100° C), and is very salty. The important gradient zone separates these zones. The gradient zone acts as a transparent insulator, permitting the sunlight to be trapped in the hot bottom layer (from which useful heat is withdrawn). This is because the salt gradient, which increases the brine density with depth, counteracts the buoyancy effect of the warmer water below (which would otherwise rise to the surface and lose its heat to the air). An organic Rankine cycle engine is used to convert the thermal energy to electricity.

- **sample (coal)**

A representative fraction of a coal bed collected by approved methods, guarded against contamination or adulteration, and analyzed to determine the nature; chemical, mineralogic, and (or) petrographic composition; percentage or parts-per-million content of specified constituents; heat value; and possibly the reactivity of the coal or its constituents.

- **schedule**

A statement of the pricing format of electricity and the terms and conditions governing its applications.

- **scheduled outage**

The shutdown of a generating unit, transmission line, or other facility for inspection or maintenance, in accordance with an advance schedule.

- **scheduling coordinators**

Entities certified by the Federal Energy Regulatory Commission (FERC) that act on behalf of generators, supply aggregators (wholesale marketers), retailers, and customers to schedule the distribution of electricity.

- **scoop loading**

An underground loading method by which coal is removed from the working face by a tractor unit equipped with a hydraulically operated bucket attached to the front; also called a front-end loader.

- **screenings**

The undersized coal from a screening process, usually one-half inch or smaller.

seam

A bed of coal lying between a roof and floor. Equivalent term to bed, commonly used by industry.

seasonal energy efficiency ratio (seer)

Ratio of the cooling output divided by the power consumption. It is the Btu of cooling output during its normal annual usage divided by the total electric energy input in watt hours during the same period. This is a measure of the cooling performance for rating central air conditioners and central heat pumps. The appliance standards required a minimum SEER of 10 for split-system central air conditioners and for split-system central heat pumps in 1992. (The average heat pump or central air conditioner sold in 1986 had an SEER of about 9.)

seasonal pricing

A special electric rate feature under which the price per unit of energy depends on the season of the year.

seasonal rates

Different seasons of the year are structured into an electric rate schedule whereby an electric utility provides service to consumers at different rates. The electric rate schedule usually takes into account demand based on weather and other factors.

seasonal units

Housing units intended for occupancy at only certain seasons of the year. Seasonal units include units intended only for recreational use, such as beach cottages and hunting cabins. It is not likely that this type of unit will be the usual residence for a household, because it may not be fit for living quarters for more than half of the year.

seasoned wood

Wood, used for fuel, that has been air dried so that it contains 15 to 20 percent moisture content.

secondary heating equipment

Space-heating equipment used less often than the main space-heating equipment.

secondary heating fuel

Fuels used in secondary space-heating equipment.

securitization

A proposal for issuing bonds that would be used to buy down existing power contracts or other obligations. The bonds would be repaid by designating a portion of future customer bill payments. Customer bills would be lowered, since the cost of bond payments would be less than the power contract costs that would be avoided.

securitize

To aggregate contracts into one pool, which then offers shares for sale in the investment market. This strategy diversifies project risks from what they would be if each project were financed individually, thereby reducing the cost of financing.

- **selective absorber**
  A solar absorber surface that has high absorbtance at wavelengths corresponding to that of the solar spectrum and low emittance in the infrared range.

- **self-generator**
  A plant whose primary product is not electric power, but does generate electricity for its own use or for sale on the grid; for example, industrial combined heat and power plants.

- **seller type**
  Categories of major refiners and other refiners and gas plant operators.

- **semiconductor**
  Any material that has a limited capacity for conducting an electric current. Certain semiconductors, including silicon, gallium arsenide, copper indium diselenide, and cadmium telluride, are uniquely suited to the photovoltaic conversion process.

- **separate metering**
  Measurement of electricity or natural gas consumption in a building using a separate meter for each of several tenants or establishments in the building.

- **separative work unit (swu)**
  The standard measure of enrichment services. The effort expended in separating a mass F of feed of assay x_f into a mass P of product assay x_p and waste of mass W and assay x_w is expressed in terms of the number of separative work units needed, given by the expression $SWU = WV(x_w) + PV(x_p) - FV(x_f)$, where $V(x)$ is the "value function," defined as $V(x) = (1 - 2x) \ln((1 - x)/x)$.

- **septic tank**
  A tank in which the solid matter of continuously flowing sewage is disintegrated by bacteria.

- **series connection**
  A way of joining photovoltaic cells by connecting positive leads to negative leads; such a configuration increases the voltage.

- **series resistance**
  Parasitic resistance to current flow in a cell due to mechanisms such as resistance from the bulk of the semiconductor material, metallic contacts, and interconnections.

- **service area**
  The territory in which a utility system or distributor is authorized to provide service to consumers.

- **service well**
  A well drilled, completed, or converted for the purpose of supporting production in an

existing field. Wells of this class also are drilled or converted for the following specific purposes: gas injection (natural gas, propane, butane or fuel-gas); water injection; steam injection; air injection; salt water disposal; water supply for injection; observation; and injection for in-situ combustion.

■ **shaft mine**
A mine that reaches the coal bed by means of a vertical shaft.

■ **shakes/shingles**
Flat pieces of weatherproof material laid with others in a series of overlapping rows as covering for roofs and sometimes the sides of buildings. Shakes are similar to wood shingles, but instead of having a cut and smoothly planed surface, shakes have textured grooves and a rough or "split" appearance to give a rustic feeling.

■ **shallow pitting**
Testing a potential mineral deposit by systematically sinking small shafts into the earth and analyzing the material recovered.

■ **shell storage capacity**
The design capacity of a petroleum storage tank which is always greater than or equal to working storage capacity.

■ **short circuit**
An electric current taking a shorter or different path than intended.

■ **short circuit current**
The current flowing freely through an external circuit that has no load or resistance; the maximum current possible.

■ **short purchases**
A single shipment of fuel or volumes of fuel purchased for delivery within 1 year. Spot purchases are often made by a user to fulfill a certain portion of energy requirements, to meet unanticipated energy needs, or to take advantage of low-fuel prices.

■ **short term sales**
Any short-term purchase covering a time period of 2 years or less. Purchases from intrastate pipelines pursuant to Section 311(b) of the NGPA of 1978 are classified as short-term sales, regardless of the stated contract term.

■ **short ton**
A unit of weight equal to 2,000 pounds.

■ **short-term debt or borrowings**
Debt securities or borrowings having a maturity of less than one year.

■ **short-term purchase**
A purchase contract under which all deliveries of materials are scheduled to be completed by the end of the first calendar year following the contract-signing year. Deliveries can be made during the contract year, but

deliveries are not scheduled to occur beyond the first calendar year thereafter.

- **shortwall mining**
A form of underground mining that involves the use of a continuous mining machine and movable roof supports to shear coal panels 150 to 200 feet wide and more than half a mile long. Although similar to long wall mining, shortwall mining is generally more flexible because of the smaller working area. Productivity is lower than with long wall mining because the coal is hauled to the mine face by shuttle cars as opposed to conveyors.

- **shrinkage**
The volume of natural gas that is transformed into liquid products during processing, primarily at natural gas liquids processing plants.

- **shut in**
Closed temporarily; wells and mines capable of production may be shut in for repair, cleaning, inaccessibility to a market, etc.

- **shutdown date**
Month and year of shutdown for fuel discharge and refueling. The date should be the point at which the reactor became subcritical.

- **shut-in royalty**
A royalty paid by a lessee as compensation for a lessor's loss of income because the lessee has deferred production from a property that is known to be capable of producing minerals. Shut in may be caused by a lack of a ready market, by a lack of transportation facilities, or by other reasons. A shut-in royalty may or may not be recoverable out of future production.

- **sidetrack drilling**
This is a remedial operation that results in the creation of a new section of well bore for the purpose of (1) detouring around junk, (2) redrilling lost holes, or (3) straightening key seats and crooked holes. Directional "side-track" wells do not include footage in the common bore that is reported as footage for the original well.

- **siding**
An exterior wall covering material made of wood, plastic (including vinyl), or metal. Siding is generally produced in the shape of boards and is applied to the outside of a building in overlapping rows.

- **silicon**
A semiconductor material made from silica, purified for photovoltaic applications.

- **silt**
Waste from Pennsylvania anthracite preparation plants, consisting of coarse rock fragments containing as much as 30 percent small-sized coal; sometimes defined as including

very fine coal particles called silt. Its heat value ranges from 8 to 17 million Btu per short ton.

- **silt, culm, refuse bank, or slurry dam mining**

  A mining operation producing coal from these sources of coal.

- **single crystal silicon (czochralsky)**

  Silicon cells with a well-ordered crystalline structure consisting of one crystal (usually obtained by means of the Czochralsky growth technique and involving ingot slicing), composing a module. Ribbon silicon is excluded.

- **single crystal silicon**

  An extremely pure form of crystalline silicon produced by dipping a single crystal seed into a pool of molten silicon under high vacuum conditions and slowly withdrawing a solidifying single crystal boule (rod) of silicon. The boule is sawed into thin silicon wafers and fabricated into single-crystal photovoltaic cells.

- **single purpose project**

  A hydroelectric project constructed only to generate electricity.

- **single-circuit line**

  A transmission line with one electric circuit. For three-phase supply, a single circuit requires at least three conductors, one per phase.

- **sinter**

  A chemical sedimentary rock deposited by precipitation from mineral waters, especially siliceous sinter and calcareous sinter.

- **site characterization**

  An onsite investigation at a known or suspected contaminated waste or release site to determine the extent and type(s) of contamination.

- **site energy consumption**

  The Btu value of energy at the point it enters the home, building, or establishment, sometimes referred to as "delivered" energy.

- **site energy**

  The Btu value of energy at the point it enters the home, sometimes referred to as "delivered" energy. The site value of energy is used for all fuels, including electricity.

- **site-specific information dsm program assistance**

  A DSM (demand-side management) assistance program that provides quidance on energy efficiency and load management options tailored to a particular customer' s facility; it often involves an on-site inspection of the customer facility to identify cost-effective DSM actions that could be taken. They include audits, engineering design calculations on information provided about the building, and technical assistance to architects and engineers who design new facilities.

- **slope mine**

A mine that reaches the coal bed by means of an inclined opening.

- **slot**

A physical position in a rack in a storage pool that is intended to be occupied by an intact assembly or equivalent (that is, a canister or an assembly skeleton).

- **sludge**

A dense, slushy, liquid-to-semi fluid product that accumulates as an end result of an industrial or technological process designed to purify a substance. Industrial sludges are produced from the processing of energy-related raw materials, chemical products, water, mined ores, sewerage, and other natural and man-made products. Sludges can also form from natural processes, such as the run off produced by rain fall, and accumulate on the bottom of bogs, streams, lakes, and tidelands.

- **slurry**

A viscous liquid with a high solids content.

- **slurry dam**

A repository for the silt or culm from a preparation plant.

- **small pickup truck**

A pickup truck weighing under 4,500 lbs GVW.

- **small power producer (spp)**

Under the Public Utility Regulatory Policies Act (PURPA), a small power production facility (or small power producer) generates electricity using waste, renewable (biomass, conventional hydroelectric, wind and solar, and geothermal) energy as a primary energy source. Fossil fuels can be used, but renewable resource must provide at least 75 percent of the total energy input.

- **sodium lights**

A type of high intensity discharge light that has the most lumens per watt of any light source.

- **sodium silicate**

A grey-white powder soluble in alkali and water, insoluble in alcohol and acid. Used to fireproof textiles, in petroleum refining and corrugated paperboard manufacture, and as an egg preservative. Also referred to as liquid gas, silicate of soda, sodium metasilicate, soluble glass, and water glass.

- **sodium tripolyphosphate**

A white powder used for water softening and as a food additive and texturizer.

- **solar cells or photovoltaic energy**

We can also change the sunlight directly to electricity using solar cells.

Solar cells are also called photovoltaic cells - or PV cells for short - and can be found on many small appliances, like calculators, and even on spacecraft. They were first developed in the 1950s for use on U.S. space satellites. They are made of silicon, a special type of melted sand.

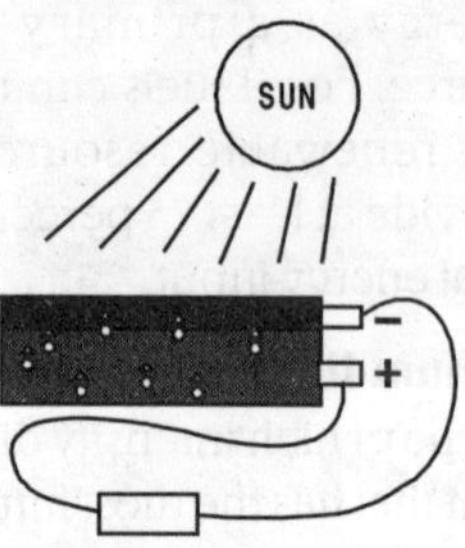

When sunlight strikes the solar cell, electrons (red circles) are knocked loose. They move toward the treated front surface (dark blue color). An electron imbalance is created between the front and back. When the two surfaces are joined by a connector, like a wire, a current of electricity occurs between the negative and positive sides.

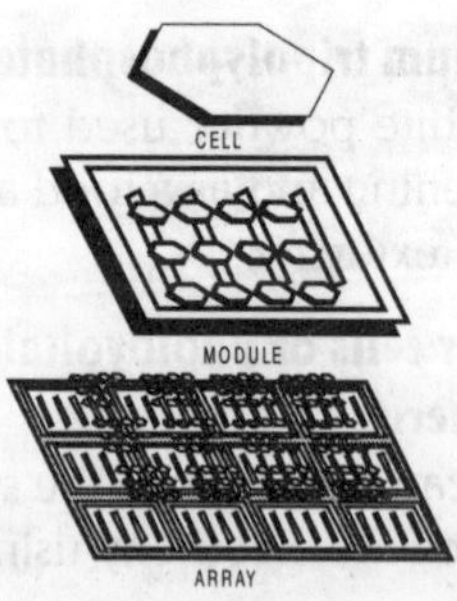

- **solar constant**

The average amount of solar radiation that reaches the earth's upper atmosphere on a surface perpendicular to the sun's rays; equal to 1353 Watts per square meter or 492 Btu per square foot.

- **solar cooling**

The use of solar thermal energy or solar electricity to power a cooling appliance. There are five basic types of solar cooling technologies: absorption cooling, which can use solar thermal energy to vaporize the refrigerant; desiccant cooling, which can use solar thermal energy to regenerate (dry) the desiccant; vapor compression cooling, which can use solar thermal energy to operate a Rankine-cycle heat engine; and evaporative coolers ("swamp" coolers), and heat-pumps and air conditioners that can by powered by solar photovoltaic systems.

- **solar declination**

The apparent angle of the sun north or south of the earth's equatorial plane. The earth's rotation on its axis causes a daily change in the declination.

- **solar energy**

The radiant energy of the sun, which can be converted into other forms of energy, such as heat or electricity. We have always used the energy of the sun as far back as humans have existed on this planet. As far back as 5,000 years ago, people "worshipped" the sun. Ra, the sun-god, who was considered the first king of Egypt. In Mesopotamia, the sun-god Shamash was a major deity and was equated with justice. In Greece there were two sun deities, Apollo and Helios. The influence of the sun also appears in other religions - Zoroastrianism, Mithraism, Roman religion, Hinduism, Buddhism, the Druids of England, the Aztecs of Mexico, the Incas of Peru, and many Native American tribes.

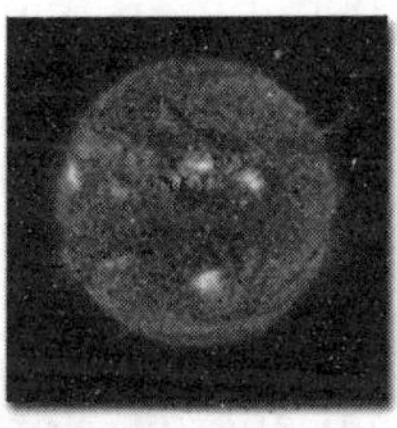

- **solar hot water**

In the 1890s solar water heaters were being used all over the United States. They proved to be a big improvement over wood and coal-burning stoves. Artificial gas made from coal was available too to heat water, but it cost 10 times the price we pay for natural gas today. And electricity was even more expensive if you even had any in your town!

Many homes used solar water heaters. In 1897, 30 percent of the homes in Pasadena, just east of Los Angeles, were equipped with solar water heaters. As mechanical improvements were made, solar systems were used in Arizona, Florida and many other sunny parts of the United States. The picture shown here is a solar water heater installed on the front roof of a house in Pomona Valley, California, in 1911 (the panels are circled above the four windows).

- **solar pond**

A body of water that contains brackish (highly saline) water that forms layers of differing salinity (stratifies) that absorb and trap solar energy. Solar ponds can be used to provide heat for industrial or agricultural processes, building heating and cooling, and to generate electricity.

- **solar power satellites**

One suggestion for energy in the future is to put huge solar power satellites into orbit around the earth. They would collect solar energy from the sun, convert it to electricity and beam it to Earth as microwaves or some other form of transmission. The power would have no greenhouse gas emissions, but microwave beams might affect health adversely. And frequent rocket launches may harm the upper atmosphere. This idea may not be practical for another century; if at all.

The picture on the right is an early and simple drawing of how a space solar power satellite would beam energy to electrical power grid on Earth.

- **solar power tower**

A solar energy conversion system that uses a large field of independently adjustable mirrors (heliostats) to focus solar rays on a near single point atop a fixed tower (receiver). The concentrated energy may be used to directly heat the working fluid of a Rankine cycle engine or to heat an intermediary thermal storage medium (such as a molten salt).

- **solar radiation**

A general term for the visible and near visible (ultraviolet and near-infrared) electromagnetic radiation that is emitted by the sun. It has a spectral, or wavelength, distribution that corresponds to different energy levels; short wavelength radiation has a higher energy than long-wavelength radiation.

- **solar spectrum**

The total distribution of electromagnetic radiation emanating from the sun. The different regions of the solar spectrum are described by their wavelength range. The visible region extends from about 390 to 780 nanometers (a nanometer is one billionth of one meter). About 99 percent of solar radiation is contained in a wavelength region from 300 nm (ultraviolet) to 3,000 nm (near-infrared). The combined radiation in the wavelength region from 280 nm to 4,000 nm is called the broadband, or total, solar radiation.

- **solar thermal collector**

A device designed to receive solar radiation and convert it to thermal energy. Normally, a solar thermal collector includes a frame, glazing, and an absorber, together with appropriate insulation. The heat collected by the solar collector may be used immediately or stored for later use. Solar collectors are used for space heating; domestic hot water heating; and heating swimming pools, hot tubs, or spas.

- **solar thermal collector, high temperature**

A collector that generally operates at temperatures above 180 degrees Fahrenheit.

- **solar thermal collector, low-temperature**

A collector that generally operates at temperatures below 110 degrees Fahrenheit. Typically, it has no glazing or insulation and is made of plastic or rubber, although some are made of metal.

- **solar thermal collector, medium-temperature**

A collector that generally operates at temperatures of 140 degrees F to 180 degrees Fahrenheit, but can also operate at temperatures as low as 110 degrees Fahrenheit. Typically, it has one or two glazings, a metal frame, a metal absorption panel with integral flow channels or attached tubing (liquid collector) or with integral ducting (air collector) and insulation on the sides and back of the panel.

- **solar thermal collector, special**

An evacuated tube collector or a concentrating (focusing) collector. Special collectors operate in the temperature range from just above ambient temperature (low concentration for pool heating) to several hundred degrees Fahrenheit (high concentration for air conditioning and specialized industrial processes).

- **solar thermal electricity**

Solar energy can also be used to make electricity.

Some solar power plants, like the one in the picture to the right in California's Mojave Desert, use a highly curved mirror called a parabolic trough to focus the sunlight on a pipe running down a central point above the curve of the mirror. The mirror focusses the sunlight to strike the pipe, and it gets so hot that it can boil water into steam. That steam can then be used to turn a turbine to make electricity.

In California's Mojave desert, there are huge rows of solar mirrors arranged in what's called "solar thermal power plants" that use this idea to make electricity for more than 350,000 homes. The problem with solar energy is that it works only when the sun is shining. So, on cloudy days and at night, the power plants can't create energy. Some solar plants, are a "hybrid" technology. During the daytime they use the sun. At night and on cloudy days they burn natural gas to boil the water so they can continue to make electricity.

Another form of solar power plants to make electricity is called a Central Tower Power Plant, like the one to the right - the Solar Two Project.

- **solar thermal panels**

A system that actively concentrates thermal energy from the sun by means of solar collector panels. The panels typically consist of fat, sun-oriented boxes with transparent covers, containing water tubes of air baffles under a blackened heat absorbent panel. The energy is usually used for space heating, for water heating, and for heating swimming pools.

- **solar thermal parabolic dishes**

A solar thermal technology that uses a modular mirror system that approximates a parabola and incorporates two-axis tracking to focus the sunlight onto receivers located at the focal point of each dish. The mirror system typically is made from a number of mirror facets, either glass or polymer mirror, or can consist of a single stretched membrane using a polymer mirror. The concentrated sunlight may be used directly by a Stirling, Rankine, or Brayton cycle heat engine at the focal point of the receiver or to heat a working fluid that is piped to a central engine. The primary applications include remote electrification, water pumping, and grid-connected generation.

- **source material**

The term "source material" means (1) uranium, thorium, or any other material that is determined by the Atomic Energy Commission pursuant to the provisions of section 61 of the Atomic Energy Act of 1954, as amended, to be source material; or (2) ores containing one or more of the foregoing materials, in such concentration as the Commission may by regulation determine from time to time.

- **space heating**

The use of energy to generate heat for warmth in housing units using space-heating equipment. The equipment could be the main space-heating equipment or secondary space-heating equipment. It does not include the use of energy to operate appliances (such as lights, televisions, and refrigerators) that give off heat as a byproduct.

- **special collector**

An evacuated tube collector or a concentrating (focusing) collector. Special collectors operate in the temperature range from just above ambient temperature (low concentration for pool heating) to several hundred degrees Fahrenheit (high concentration for air conditioning and specialized industrial processes).

- **special contract rate schedule**
  An electric rate schedule for an electric service agreement between a utility and another party in addition to, or independent of, any standard rate schedule.

- **special naphthas**
  All finished products within the naphtha boiling range that are used as paint thinners, cleaners, or solvents. These products are refined to a specified flash point. Special naphthas include all commercial hexane and cleaning solvents conforming to ASTM Specification D1836 and D484, respectively. Naphthas to be blended or marketed as motor gasoline or aviation gasoline, or that are to be used as petrochemical and synthetic natural gas (SNG) feed stocks are excluded.

- **special nuclear material**
  The term "special nuclear material" means (1) plutonium, uranium enriched in the isotope 233 or in the isotope 235, and any other material that the Atomic Energy Commission, pursuant to the provisions of section 51 of the Atomic Energy Act of 1954, as amended, determines to be special nuclear material, but does not include source material; or (2) any material artificially enriched by any of the foregoing, but does not include source material.

- **special purpose rate schedule**
  An electric rate schedule limited in its application to some particular purpose or process within one, or more than one, type of industry or business.

- **specular reflectors**
  Specular reflectors have mirror like characteristics (the word "specular" is derived from the Greek word meaning mirror). The most common materials used for ballasts, the devices that turn on and operate Fluorescent tubes, are aluminum and silver. Silver has the highest reflectivity; aluminum has the lowest cost. The materials and shape of the reflector are designed to reduce absorption of light within the fixture while delivering light in the desired angular pattern. Adding (or retrofitting) specular reflectors to an existing light fixture is frequently implemented as a conservation measure.

- **speculative resources (coal)**
  Undiscovered coal in beds that may occur either in known types of deposits in a favorable geologic setting where no discoveries have been made, or in deposits that remain to be recognized. Exploration that confirms their existence and better defines their quantity and quality would permit their reclassification as identified resources.

speculative resources (uranium)

Uranium in addition to Estimated Additional Resources (EAR) that is thought to exist, mostly on the basis of indirect evidence and geological extrapolations, in deposits discoverable with existing exploration techniques. The locations of deposits in this category can generally be specified only as being somewhere within given regions or geological trends. The existence and size of such deposits are speculative. The estimates in this category are less reliable than estimates of EAR. SR corresponds to DOE's Possible Potential Resources plus Speculative Potential Resources categories.

spending on pollution control in USA

Capital expenditure of United States refining industries on pollution-abatement measures have accounted for a varying share of total capital investment which was highest after the Clean Air Act amendments of 1990.

In three years –1992, 1993, and 1994 – they accounted for over half of all capital expenditure and reached nearly $3,000 million.

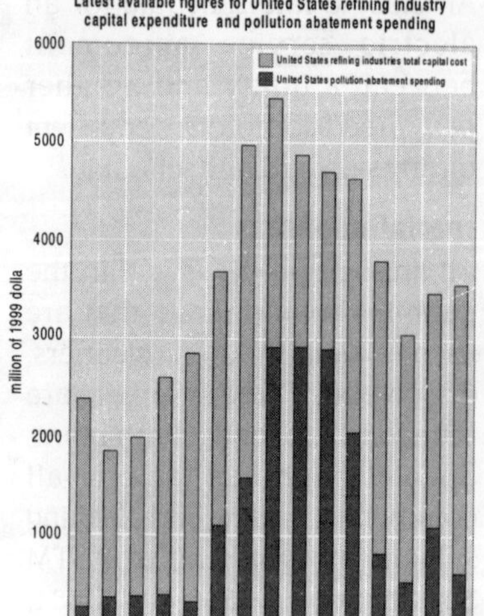

spent fuel disassembly hardware

The skeleton of a fuel assembly after the fuel rods have been removed. Generally, SFD hardware for PWR assemblies includes guide tubes; instrument tubes, top and bottom nozzles; grid spacers; hold-down springs; and attachment components, such as nuts and locking caps. For BWR fuel assemblies, SFD hardware includes the top and bottom tie plates, compression springs for individual fuel rods, grid spacers, and water rods.

spent fuel

Irradiated fuel that is permanently discharged from a reactor. Except for possible reprocessing, this fuel must eventually be removed from its temporary storage location at the reactor site and placed in a permanent repository. Spent fuel is typically measured either in metric tons of heavy metal (i.e., only the heavy metal content of the spent fuel is considered) or in metric tons of initial heavy metal (essentially, the initial mass of the fuel before irradiation). The difference between these two quantities is the weight of the fission products.

- **spent liquor**
The liquid residue left after an industrial process; can be a component of waste materials used as fuel.

- **spillway**
A passage for surplus water to flow over or around a dam.

- **spinning reserve**
That reserve generating capacity running at a zero load and synchronized to the electric system.

- **split system**
When applied to electric air-conditioning equipment, it means a two-part system—an indoor unit and an outdoor unit. The indoor unit is an evaporator coil mounted in the indoor circulating air system, and the outdoor unit is an air-cooled condensing unit containing an electric motor-driven compressor, a condenser fan and a fan motor.

- **split tails**
Use of one tails assay for transaction of enrichment services and a different tails assay for operation of the enrichment plant. This mode of operations typically increases the use of uranium, which is relatively inexpensive, while decreasing the use of separative work, which is expensive.

- **spontaneous combustion, or self-heating, of coal**
A naturally occurring process caused by the oxidation of coal. It is most common in low-rank coals and is a potential problem in storing and transporting coal for extended periods. Factors involved in spontaneous combustion include the size of the coal (the smaller sizes are more susceptible), the moisture content, and the sulfur content. Heat buildup in stored coal can degrade the quality of coal, cause it to smolder, and lead to a fire.

- **spot market (natural gas)**
A market in which natural gas is bought and sold for immediate or very near-term delivery, usually for a period of 30 days or less. The transaction does not imply a continuing arrangement between the buyer and the seller. A spot market is more likely to develop at a location with numerous pipeline interconnections, thus allowing for a large number of

buyers and sellers. The Henry Hub in southern Louisiana is the best known spot market for natural gas.

- **spot market (uranium)**
Buying and selling of uranium for immediate or very near-term delivery. It typically involves transactions for delivery of up to 500,000 pounds U_3O_8 within a year of contract execution.

- **spot price**
The price for a one-time open market transaction for immediate delivery of a specific quantity of product at a specific location where the commodity is purchased "on the spot" at current market rates.

- **spot purchases**
A single shipment of fuel or volumes of fuel purchased for delivery within 1 year. Spot purchases are often made by a user to fulfill a certain portion of energy requirements, to meet unanticipated energy needs, or to take advantage of low-fuel prices.

- **stability**
The property of a system or element by virture of which its output will ultimately attain a steady state. The amount of power that can be transferred from one machine to another following a disturbance. The stability of a power system is its ability to develop restoring forces equal to or greater than the disturbing forces so as to maintain a state of equilibrium.

- **stabilization lagoon**
A shallow artificial pond used for the treatment of wastewater. Treatment includes removal of solid material through sedimentation, the decomposition of organic material by bacteria, and the removal of nutrients by algae.

- **stack**
A tall, vertical structure containing one or more flues used to discharge products of combustion to the atmosphere.

- **stand-alone generator**
A power source/generator that operates independently of or is not connected to an electric transmission and distribution network; used to meet a load(s) physically close to the generator.

- **standard contract**
The agreement between the Department of Energy (DOE) and the owners or generators of spent nuclear fuel and high-level radioactive waste, under which DOE will make available nuclear waste disposal services to those owners and generators.

- **standard fluorescent**
A light bulb made of a glass tube coated on the inside with fluorescent material, which produces light by passing

electricity through mercury vapor causing the fluorescent coating to glow or fluoresce.

- **standard industrial classification (sic)**

Replaced with North American Industry Classification System.

- **standby charge**

A charge for the potential use of a utility service, usually done by an agreement with another electric utility service. These services include system backup support and other running and quick-start capabilities.

- **standby electricity generation**

Involves use of generators during times of high demand on utilities to avoid extra "peak-demand" charges.

- **standby facility**

A facility that supports a utility system and is generally running under no-load. It is available to replace or supplement a facility normally in service.

- **standby heat loss**

A term used to describe heat energy lost from a water heater tank.

- **standby service**

Support service that is available as needed to supplement a customer, a utility system, or another utility if a schedule or an agreement authorizes the transaction. The service is not regularly used.

- **startup test phase of nuclear power plant**

A nuclear power plant that has been licensed by the Nuclear Regulatory Commission to operate but is still in the initial testing phase, during which the production of electricity may not be continuous. In general, when the electric utility is satisfied with the plant's performance, it formally accepts the plant from the manufacturer and places it in commercial operation status. A request is then submitted to the appropriate utility rate commission to include the power plant in the rate base calculation.

- **startup/flame stabilization fuel**

Any fuel used to initiate or sustain combustion or used to stabilize the height of flames once combustion is underway.

- **state**

One of the 50 States, including adjacent outer continental shelf areas, or the District of Columbia.

- **state permit/license/mine number**

Code assigned to a mining operation by the state in which the operation is located.

- **state severance taxes**

Any severance, production, or similar tax, fee, or other levy imposed on the production of crude oil, natural gas, or coal by any State, local government acting under authority of State law, or by an Indian tribe recognized as eligible for services by the Secretary of the Interior.

- **station (electric)**
A plant containing prime movers, electric generators, and auxiliary equipment for converting mechanical, chemical, and/or nuclear energy into electric energy.

- **station use**
Energy that is used to operate an electric generating plant. It includes energy consumed for plant lighting, power, and auxiliary facilities, regardless of whether the energy is produced at the plant or comes from another source.

- **steam (purchased)**
Steam, purchased for use by a refinery, that was not generated from within the refinery complex.

- **steam boiler**
A type of furnace in which fuel is burned and the heat is used to produce steam.

- **steam coal**
All nonmetallurgical coal.

- **steam electric power plant (conventional)**
A plant in which the prime mover is a steam turbine. The steam used to drive the turbine is produced in a boiler where fossil fuels are burned.

- **steam expenses**
The cost of labor, materials, fuel, and other expenses incurred in production of steam for electric generation.

- **steam for heating/cooling**
Steam produced at a combined heat and power plant for the purpose of heating and/or cooling space, such as district heating systems.

- **steam from other sources**
Steam purchased, transferred from another department of the utility, or acquired from others under. a joint-facility operating agreement.

- **steam or hot water radiators or baseboards**
A distribution system where steam or hot water circulates through cast-iron radiators or baseboards. some other types of equipment in the building may be used to produce the steam or hot water or it may enter the building already heated as part of a district hot water system. hot water does not include domestic hot water used for cooking and cleaning.

- **steam or hot-water system**
Either of two types of a central space-heating system that supplies steam or hot water to radiators, convectors, or pipes.

The more common type supplies either steam or hot water to conventional radiators, baseboard radiators, convectors, heating pipes embedded in the walls or ceilings, or heating coils or equipment that are part of a combined heating/ventilating or heating/air-conditioning system. The other type supplies radiant heat through pipes that carry hot water and are held in a concrete slab floor.

- **steam transferred-credit**

The expenses of producing steam are charged to others or to other utility departments under a joint operating arrangement.

- **steam turbine**

A device that converts high-pressure steam, produced in a boiler, into mechanical energy that can then be used to produce electricity by forcing blades in a cylinder to rotate and turn a generator shaft.

- **steam**

Water in vapor form; used as the working fluid in steam turbines and heating systems.

- **still gas (refinery gas)**

Any form or mixture of gases produced in refineries by distillation, cracking, reforming, and other processes. The principal constituents are methane, ethane, ethylene, normal butane, butylene, propane, propylene, etc. Still gas is used as a refinery fuel and a petrochemical feedstock. The conversion factor is 6 million BTU's per fuel oil equivalent barrel.

- **stock change**

The difference between stocks at the beginning of the reporting period and stocks at the end of the reporting period. Note: A negative number indicates a decrease (i.e., a draw down) in stocks and a positive number indicates an increase (i.e., a buildup) in stocks during the reporting period.

- **stocks**

Inventories of fuel stored for future use.

- **storage additions**

Volumes of gas injected or otherwise added to underground natural gas reservoirs or liquefied natural gas storage.

- **storage agreement**

Any contractual arrangement between the responding company and a storage operator under which gas was stored for, or gas storage service was provided to, the responding company by the storage operator, irrespective of any responding company ownership interest in either the storage facilities or stored gas.

- **storage capacity**

The amount of energy an energy storage device or system can store.

■ **storage field capacity (underground gas storage)**
The presently developed maximum capacity of a field.

■ **storage hydroelectric plant**
A hydroelectric plant with reservoir storage capacity for power use.

■ **storage site**
Spent nuclear fuel storage pool or dry cask storage facility, usually located at the reactor site, as licensed by (or proposed to be licensed by) the Nuclear Regulatory Commission (NRC).

■ **storage withdrawals**
Total volume of gas withdrawn from underground storage or from liquefied natural gas storage over a specified amount of time.

■ **storm door**
A second door installed outside or inside a prime door creating an insulating air space. Included are sliding glass doors made of double glass or of insulating glass such as thermopane and sliding glass doors with glass or Plexiglas placed on either the outside or inside of the door to create an insulating air space. Not included are doors or sliding glass doors covered by plastic sheets or doors with storm window covering on just the glass portion of the door.

■ **storm or multiple glazing**
A building shell conservation feature consisting of storm windows, storm doors, or double- or triple-paned glass that are placed on the exterior of the building to reduce the rate of heat loss.

■ **storm window**
A window or glazing material placed outside or inside a window creating an insulating air space. Plastic material over windows is counted a a storm window if the same plastic material can be used year after year or if the plastic is left in place year-round and is in good condition (no holes or tears). If the plastic material must be put up new each year, it is not counted as a storm window. It is counted as "plastic coverings." Glass or Plexiglas placed over windows on either the interior or exterior side is counted as storm windows.

■ **stranded benefits**
Benefits associated with regulated retail electric service which may be at risk under open market retail competition. Examples include conservation programs, fuel diversity, reliability of supply, and tax revenues based on utility revenues.

■ **stranded costs**
Costs incurred by a utility which may not be recoverable under market-based retail competition. Examples include undepreciated generating facilities, deferred costs, and long-term contract costs.

strategic petroleum reserve (spr)
Petroleum stocks maintained by the Federal Government for use during periods of major supply interruption.

stratigraphic test well
A geologically directed drilling effort to obtain information pertaining to a specific geological condition that might lead toward the discovery of an accumulation of hydrocarbons. Such wells are customarily drilled without the intention of being completed for hydrocarbon production. This classification also includes tests identified as core tests and all types of expendable holes related to hydrocarbon exploration.

stratosphere
The region of the upper atmosphere extending from the tropopause (8 to 15 kilometers altitude) to about 50 kilometers. Its thermal structure, which is determined by its radiation balance, is generally very stable with low humidity.

stream-flow
The rate at which water passes a given point in a stream, usually expressed in cubic feet per second.

strip mine
An open cut in which the overburden is removed from a coal bed prior to the removal of coal.

strip mining (surface)
A method used on flat terrain to recover coal by mining long strips successively; the material excavated from the strip being mined is deposited in the strip previously mined.

strip or stripping ratio
The amount of overburden that must be removed to gain access to a unit amount of coal. A stripping ratio may be expressed as (1) thickness of overburden to thickness of coal, (2) volume of overburden to volume coal, (3) weight of overburden to weight of coal, or (4) cubic yards of overburden to tons of coal. A stripping ratio commonly is used to express the maximum thickness, volume, or weight of overburden that can be profitably removed to obtain a unit amount of coal.

stripper well
An oil or gas well that produces at relatively low rates. For oil, stripper production is usually defined as production rates of between 5 and 15 barrels of oil per day. Stripper gas production would generally be anything less than 60 thousand cubic feet per day.

styrene
A colorless, toxic liquid with a strong aromatic aroma. Insoluble in water, soluble in alcohol and ether; polymerizes rapidly; can become explosive. Used to make

polymers and copolymers, polystyrene plastics, and rubber.

- **subbituminous coal**

A coal whose properties range from those of lignite to those of bituminous coal and used primarily as fuel for steam-electric power generation. It may be dull, dark brown to black, soft and crumbly, at the lower end of the range, to bright, jet black, hard, and relatively strong, at the upper end. Subbituminous coal contains 20 to 30 percent inherent moisture by weight. The heat content of subbituminous coal ranges from 17 to 24 million Btu per ton on a moist, mineral-matter-free basis. The heat content of subbituminous coal consumed in the United States averages 17 to 18 million Btu per ton, on the as-received basis (i.e., containing both inherent moisture and mineral matter).

- **subcompact/compact passenger car**

A passenger car containing less than 109 cubic feet of interior passenger and luggage volume.

- **subdivision**

A prescribed portion of a given State or other geographical region.

- **submetered data**

End-use consumption data obtained for individual appliances when a recording device has been attached to the appliance to measure the amount of energy consumed by the appliance.

- **subsidiary**

An entity directly or indirectly controlled by a parent company which owns 50% or more of its voting stock.

- **substation**

Facility equipment that switches, changes, or regulates electric voltage.

- **subtransmission**

A set of transmission lines of voltages between transmission voltages and distribution voltages. Generally, lines in the voltage range of 69 kV to 138 kV.

- **sulfur**

A yellowish nonmetallic element, sometimes known as "brimstone." It is present at various levels of concentration in many fossil fuels whose combustion releases sulfur compounds that are considered harmful to the environment. Some of the most commonly used fossil fuels are categorized according to their sulfur content, with lower sulfur fuels usually selling at a higher price.

- **sulfur dioxide (so2)**

A toxic, irritating, colorless gas soluble in water, alcohol, and ether. Used as a chemical intermediate, in paper pulping and ore refining, and as a solvent.

- **sulfur hexafluoride (sf6)**

A colorless gas soluble in alcohol and ether, and slightly less soluble in water. It is used as a dielectric in electronics. It possesses the highest 100-year Global Warming Potential of any gas (23,900).

- **sulfur oxides (sox)**

Compounds containing sulfur and oxygen, such as sulfur dioxide (SO_2) and sulfur trioxide (SO_3).

- **summer and winter peaking**

Having the annual peak demand reached both during the summer months (May through October) and during the winter months (November through April).

- **sunk cost**

Part of the capital costs actually incurred up to the date of reserves estimation minus depreciation and amortization expenses. Items such as exploration costs, land acquisition costs, and costs of financing can be included.

- **superconductivity**

The abrupt and large increase in electrical conductivity exhibited by some metals as the temperature approaches absolute zero.

- **supplemental gas**

Any gaseous substance introduced into or commingled with natural gas that increased the volume available for disposition. Such substances include, but are not limited to, propane-air, refinery gas, coke-oven gas, still gas, manufactured gas, biomass gas, or air or inserts added for Btu stabilization.

- **supplemental gaseous fuels supplies**

Synthetic natural gas, propane-air, coke oven gas, refinery gas, biomass gas, air injected for Btu stabilization, and manufactured gas commingled and distributed with natural gas.

- **supply source**

May be a single completion, a single well, a single field with one or more reservoirs, several fields under a single gas-purchase contract, miscellaneous fields, a processing plant, or a field area; provided, however, that the geographic area encompassed by a single supply source may not be larger than the state in which the reserves are reported.

- **supply**

The components of petroleum supply are field production, refinery production, imports, and net receipts when calculated on a PAD District basis.

- **supply, petroleum**

A set of categories used to account for how crude oil and petroleum products are transferred, distributed, or placed into the supply stream. The categories include field production, refinery production, and imports. Net receipts are also included on a Petroleum Administration for Defense (PAD) District basis to account for shipments of crude oil and petroleum products across districts.

- **support equipment and facilities**

These include, but are not limited to, seismic equipment, drilling equipment, construction and grading equipment, vehicles, repair shops, warehouses, supply points, camps, and division, district, or field offices.

- **supporting structure**

The main supporting unit (usually a pole or tower) for transmission line conductors, insulators, and other auxiliary line equipment.

- **surface drilling expenses (uranium)**

These include drilling, drilling roads, site preparation, geological and other technical support, sampling, and drill-hole logging costs.

- **surface mine**

A coal-producing mine that is usually within a few hundred feet of the surface. Earth above or around the coal (overburden) is removed to expose the coal bed, which is then mined with surface excavation equipment, such as draglines, power shovels, bulldozers, loaders, and augers. It may also be known as an area, contour, open-pit, strip, or auger mine.

- **surface mining equipment**

An **Auger machine** is a large, horizontal drill, generally 3 feet or more in diameter and up to about 100 feet long. It can remove coal at a rate of more than 25 tons per minute.

A **bucket-wheel** excavator is a continuous digging machine equipped with a broom on which is mounted a rotating wheel with buckets along its edge. The buckets scoop up material, then empty onto a conveyor leading to a spoil bank. It is best suited for removing overburden that does not require blasting. This excavator is not widely used in the United States.

A **bulldozer** is a tractor with a movable steel blade mounted on the front. It can be used to remove overburden that needs little or no blasting.

A **carryall scraper** (or **pan scraper**) is a self-loading machine, usually self-propelled, with a scraper-like retractable bottom. It is used to excavate and haul overburden.

A **continuous surface miner**, used in some lignite mines, is equipped with crawlers, a rotating cutting head, and a conveyor. It travels over the bed, excavating a swath up to 13 feet wide and 2 feet deep.

A **dragline excavator** removes overburden to expose the coal by means of a scoop bucket that is suspended from a long boom. The dragline digs by pulling the bucket toward the machine by means of a wire rope.

A **walking dragline** is equipped with large outrigger platforms, or walking beams, instead of crawler tracks. It "walks" by the alternate movement of the walking beams.

A **drilling rig** is used to determine the amount and type of overburden overlying a coal deposit and the extent of the deposit, to delineate major geologic features, and to drill holes for explosives to fragment the overburden for easier removal.

A **front-end loader** is a tractor with a digging bucket mounted and operated on the front. It is often used to remove overburden in contour mining and to load coal.

An **hydraulic shovel** excavates and loads by means of a bucket attached to a rigid arm that is hinged to a broom.

A **power shovel** removes overburden and loads coal by means of a digging bucket mounted at the end of an arm suspended from a broom. The shovel digs by pushing the bucket forward and upward. It does not dig below the level at which it stands.

A **thin-seam miner** resembles an auger machine but has a drum-type cutting head that cuts a rectangular cross section.

■ surface mining methods

Auger mining recovers coal through the use of a large-diameter drill driven into a coal bed in the side of a surface mine pit. It usually follows contour surface mining, particularly when the overburden is too costly to excavate.

Area mining is practiced on relatively flat or gently rolling terrain and recovers coal by mining long strips successively; the material excavated from the strip being mined is deposited in the strip pit previously mined.

Contour mining is practiced when the coal is mined on hillsides. The mining follows the contour of the hillside until the overburden becomes uneconomical to remove. This method creates a shelf, or bench, on the hillside. Several variations of contour mining have been developed to control environmental problems. These methods include slope reduction (overburden is spread so that the angle of the slope on the hillside is reduced), head-of-hollow fill (overburden is placed in narrow V-shaped valleys to control erosion), and block-cut (overburden from current mining is backfilled into a previously mined cut).

Explosives casting is a technique designed to blast up to 65 percent of the overburden into the mine pit for easier removal. It differs from conventional overburden blasting, which only fractures the overburden before it is removed by excavating equipment.

Mountaintop mining, sometimes considered a variation of contour mining, refers to the mining of a coal bed that underlies the top of a mountain. The overburden, which is the mountaintop, is completely removed so that all of the coal can be recovered. The overburden material is later replaced in the mined-out area. This method leaves large plateaus of level land.

Open-pit coal mining is essentially a combination of contour and area mining methods and is used to mine thick, steeply inclined coal beds. The overburden is removed by power shovels and trucks.

- **surface rights**

Fee ownership in surface areas of land. Also used to describe a lessee's right to use as much of the surface of the land as may be reasonably necessary for the conduct of operations under the lease.

- **surplus energy**

Energy generated that is beyond the immediate needs of the producing system. This energy may be supplied by spinning reserve and sold on an interruptible basis.

- **suspended rates**

New rates that have been accepted for review by a utility commission. When these rates are suspended, they do not go into effect for a designated period of time. Charges under the new rate may be refunded after the resolution of the rate proceeding. Sustainable energy

- **swamp coolers (evaporative coolers)**

Air-conditioning equipment that removes heat by evaporating water. Evaporative cooling techniques are most commonly found in warm, dry climates such as in the Southwest, although they are found throughout the country. They usually work by spraying cool water into the air ducts, cooling the air as the spray evaporates.

- **switching station**

Facility equipment used to tie together two or more electric circuits through switches. The switches are selectively arranged to permit a circuit to be disconnected or to change the electric connection between the circuits.

- **synthetic coal**

Coal that has been processed by a coal synfuel plant; and coal-based fuels such as briquettes, pellets, or extrusions, which are formed by binding materials and processes that recycle material.

- **synthetic natural gas (sng)**
(Also referred to as substitute natural gas) A manufactured product, chemically similar in most respects to natural gas, resulting from the conversion or reforming of petroleum hydrocarbons that may easily be substituted for or interchanged with pipeline-quality natural gas.

- **system (electric)**
Physically connected generation, transmission, and distribution facilities operated as an integrated unit under one central management or operating supervision.

- **system (gas)**
An interconnected network of pipes, valves, meters, storage facilities, and auxiliary equipment used in the transportation, storage, and/or distribution of natural gas or commingled natural and supplemental gas.

- **system interconnection**
A physical connection between two electric systems that permits the transfer of electric energy in either direction.

- **tailgate**
The outlet of a natural gas processing plant where dry residue gas is delivered or re-delivered for sale or transportation.

- **tailings**
The remaining portion of a metal-bearing ore consisting of finely ground rock and process liquid after some or all of the metal, such as uranium, has been extracted.

- **tall oil**
The oily mixture of rosin acids, fatty acids, and other materials obtained by acid treatment of the alkaline liquors from the digesting (pulping) of pine wood.

- **tangible development costs**
Costs incurred during the development stage for access, mineral-handling, and support facilities having a physical nature. In mining, such costs would include tracks, lighting equipment, ventilation equipment, other equipment installed in the mine to facilitate the extraction of minerals, and supporting facilities for housing and care of work forces. In the oil and gas industry, tangible development costs would include well equipment (such as casing, tubing, pumping equipment, and well heads), as well as field storage tanks and gathering systems.

- **tank farm**
An installation used by trunk and gathering pipeline companies, crude oil producers, and terminal operators (except refineries) to store crude oil.

- **tanker and barge**
Vessels that transport crude oil or petroleum products. Data are reported for movements between PAD Districts; from a PAD District

to the Panama Canal; or from the Panama Canal to a PAD District.

- **tar sands**

Naturally occurring bitumen-impregnated sands that yield mixtures of liquid hydrocarbon and that require further processing other than mechanical blending before becoming finished petroleum products.

- **tariff**

A published volume of rate schedules and general terms and conditions under which a product or service will be supplied.

- **tax-cost**

A deduction (allowance) under U.S. Federal income taxation normally calculated under a formula whereby the adjusted basis of the mineral property is multiplied by a fraction, the numerator of which is the number of units of minerals sold during the tax year and the denominator of which is the estimated number of units of unextracted minerals remaining at the end of the tax year plus the number of units of minerals sold during the tax year.

- **temperature coefficient (of a solar photovoltaic cell)**

The amount that the voltage, current, and/or power output of a solar cell changes due to a change in the cell temperature.

- **temporarily discharged fuel**

Fuel that was irradiated in the previous fuel cycle (cycle N) and not in the following fuel cycle (cycle N+1) and that will be irradiated in a subsequent fuel cycle.

- **tennessee valley authority (tva)**

A federal agency established in 1933 to develop the Tennessee river valley region of the southeastern U.S.

- **term agreement**

Any written or unwritten agreement between two parties in which one party agrees to supply a commodity on a continuing basis to a second party for a price or for other considerations.

- **terminal location**

The physical location of one end of a transmission line segment.

- **tertiary amyl methyl ether - (ch3)2(c2h5)coch3**

An oxygenate blend stock formed by the catalytic etherification of isoamylene with methanol.

- **tertiary butyl alcohol - (ch3)3coh**

An alcohol primarily used as a chemical feedstock or a solvent or feedstock, for isobutylene production for MTBE (methyl tertiary butyl ether) and produced as a co-product of propylene oxide production or by direct hydration of isobutylene.

- **test well contribution**

A payment made to the owner of an adjacent or nearby tract who has drilled an exploratory well on that tract in exchange for information obtained from the drilling effort.

- **the carbon budget**
  - Human burning of fossil fuels and changes in landuse continue to disturb the carbon cycle, increasing the atmospheric concentration of carbon dioxide.
  - The terrestrial biosphere has historically been a source of carbon to the atmosphere; it is currently a net sink. The current terrestrial carbon sink is caused by land management practices, higher carbon dioxide, nitrogen deposition and possibly recent changes in climate.
  - The oceans, vegetation, and soils actively exchange CO2 with the atmosphere and are significant reservoirs of carbon.
  - Oceans contain about 50 times as much carbon as the atmosphere, predominantly in the form of dissolved inorganic carbon.
  - Terrestrial vegetation and soils contain about three and a half times as much carbon as the atmosphere; the exchange is controlled by photosynthesis and respiration.
  - Ocean uptake of carbon is limited by the solubility of CO2 in seawater and the slow rate of mixing between surface and deep-ocean waters.

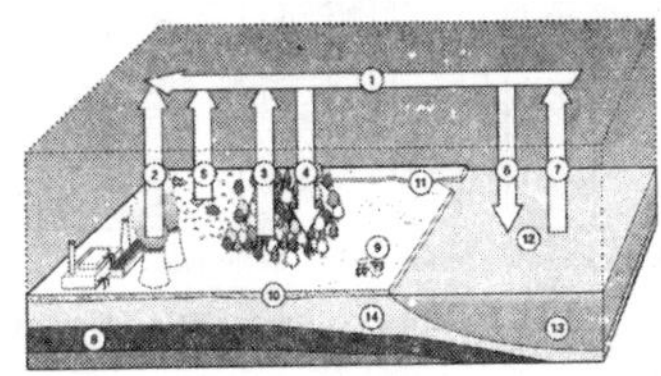

(numbers) = billions of tons of carbon transferred annually or stored

1 atmosphere (750)
2 fossil fuel burning (6)
3 respiration (60)
4 photosynthesis (62)
5 deforestation (1)
6 rainfall (92)
7 evaporation (90)
8 fossil fuels (5,000)
9 plants and animals (610)
10 soil and plant debris (1,000)
11 revers (1)
12 surface ocean (1,000)
13 intermediate/ dep ocean (38,000)
14 rocks (75,000,000)

Carbon dioxide annually emitted (+) or absorbed (–) (billions of tons of carbon)
human emission sources ('purchases'):

fossil fuel burning and cement production	+5.5 ±0.5
tropical deforestation	+1.6 ±1.0
sinks ('repayments'):	
absorption by the oceans	–2.0 ±0.8
forest regrowth in the Northern Hemisphere	–0.5 ±0.5
carbon dioxide and nitrogen fertilization, climatic effects	–1.3 ±1.5
balance ('debt') accumulates in the atmosphere):	3.3 ±0.2

the fractional distillation of oil

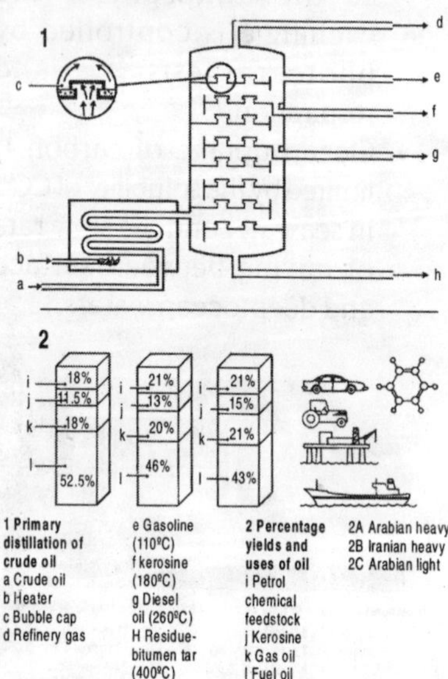

1 Primary distillation of crude oil
a Crude oil
b Heater
c Bubble cap
d Refinery gas
e Gasoline (110ºC)
f kerosine (180ºC)
g Diesel oil (260ºC)
H Residue-bitumen tar (400ºC)
2 Percentage yields and uses of oil
i Petrol chemical feedstock
j Kerosine
k Gas oil
l Fuel oil
2A Arabian heavy
2B Iranian heavy
2C Arabian light

the fuelwood crisis

- Wood provides nearly all of the energy needs of sub-Saharan African nations.
- 50–70 percent of all wood consumed on Earth is used for daily cooking purposes.
- 92 percent of all wood harvested in South Asia and 73 percent in Southeast Asia is used for fuel.
- The UN Food and Agriculture Organization (FAO) estimated that by the year 2000, some 2,400 million people were either short of fuelwood or were using it faster than it was being regenerated.
- By the year 2000 the world fuelwood deficit had reached 1,255 million cubic yards a year – the energy equivalent of 240 million tons of oil a year.
- Spectacular economies can be achieved immediately with relatively simple and inexpensive equipment.
- About 5 million fuel-saving stoves could be built for the price of one dam project in Africa.
- Use of these stoves combined would save 5 times the energy the dam could produce.

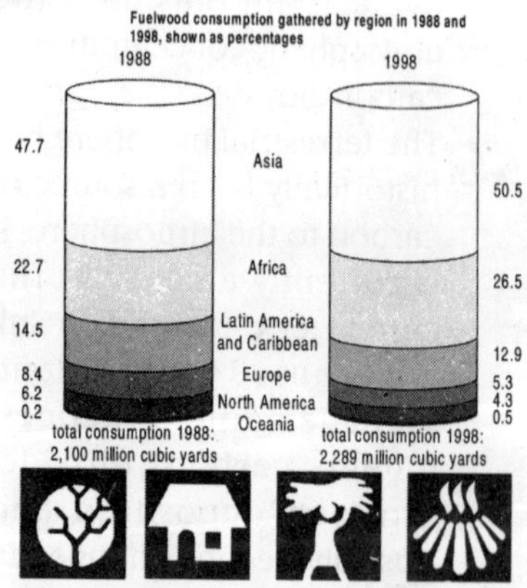

therm

One hundred thousand (100,000) Btu.

thermal

A term used to identify a type of electric generating station, capacity, capability, or output in which the source of energy for the prime mover is heat.

thermal cracking

A refining process in which heat and pressure are used to break down, rearrange, or combine hydrocarbon molecules. Thermal-cracking includes gas oil, visbreaking, fluid coking, delayed coking, and other thermal cracking processes (e.g., flexicoking).

- **thermal efficiency**
  A measure of the efficiency of converting a fuel to energy and useful work; useful work and energy output divided by higher heating value of input fuel times 100 (for percent).

- **thermal energy storage**
  The storage of heat energy during utility off-peak times at night, for use during the next day without incurring daytime peak electric rates.

- **thermal limit**
  The maximum amount of power a transmission line can carry without suffering heat-related deterioration of line equipment, particularly conductors.

- **thermal resistance (r-value)**
  This designates the resistance of a material to heat conduction. The greater the R-value the larger the number.

- **thermal storage**
  Storage of heat or heat sinks (coldness) for later heating or cooling. Examples are the storage of solar energy for night heating; the storage of summer heat for winter use; the storage of winter ice for space cooling in the summer; and the storage of electrically-generated heat or coolness when electricity is less expensive, to be released in order to avoid using electricity when the rates are higher. There are four basic types of thermal storage systems: ice storage; water storage; storage in rock, soil or other types of solid thermal mass; and storage in other materials, such as glycol (antifreeze).

- **thermocouple**
  A device consisting of two dissimilar conductors with their ends connected together. When the two junctions are at different temperatures, a small voltage is generated.

- **thermodynamics**
  A study of the transformation of energy from one form to another, and its practical application.

- **thermophotovoltaic cell**
  A device where sunlight concentrated onto a absorber heats it to a high temperature, and the thermal radiation emitted by the absorber is used as the energy source for a photovoltaic cell that is designed to maximize conversion efficiency at the wavelength of the thermal radiation.

- **thermosiphon system**
  A solar collector system for water heating in which circulation of the

collection fluid through the storage loop is provided solely by the temperature and density difference between the hot and cold fluids.

- **thermostat**

A device that adjusts the amount of heating and cooling produced and/or distributed by automatically responding to the temperature in the environment.

- **third-party dsm program sponsor**

An energy service company (ESCO) which promotes a program sponsored by a manufacturer or distributor of energy products such as lighting or refrigeration whose goal is to encourage consumers to improve energy efficiency, reduce energy costs, change the time of usage, or promote the use of a different energy source.

- **third-party transactions**

Third-party transactions are arms-length transactions between nonaffiliated firms. Producing country-to-company transactions are not considered to be third-party transactions.

- **thorium**

An element that is a byproduct of the decay of uranium.

- **three-phase power**

Power generated and transmitted from generator to load on three conductors.

- **tidal energy**

Another form of ocean energy is called tidal energy. When tides comes into the shore, they can be trapped in reservoirs behind dams. Then when the tide drops, the water behind the dam can be let out just like in a regular hydroelectric power plant.

Tidal energy has been used since about the 11th Century, when small dams were built along ocean estuaries and small streams. the tidal water behind these dams was used to turn water wheels to mill grains.

In order for tidal energy to work well, you need large increases in tides. An increase of at least 16 feet between low tide to high tide is needed. There are only a few places where this tide change occurs around the earth. Some power plants are already operating using this idea. One plant in France makes enough energy from tides (240 megawatts) to power 240,000 homes

- **tidewater piers and coastal ports (method of transportation to consumers)**

Shipments of coal moved to tidewater piers and coastal ports for further shipments to consumers via coastal water or ocean.

- **tie line**

A transmission line connecting two or more power systems.

- **time clocks or timed switches**

Time clocks are automatic controls, which turn lights off and on at predetermined times.

- **time-of-day lock-out or limit**

A special electric rate feature under which electricity usage is prohibited or restricted to a reduced level at fixed times of the day in return for a reduction in the price per kilowatt hour.

- **time-of-day pricing**

A special electric rate feature under which the price per kilowatt hour depends on the time of day.

- **time-of-day rate**

The rate charged by an electric utility for service to various classes of customers. The rate reflects the different costs of providing the service at different times of the day.

- **timing differences**

Differences between the periods in which transactions affect taxable income and the periods in which they enter into the determination of pretax accounting income. Timing differences originate in one period and reverse or "turn around" in one or more subsequent periods. Some timing differences reduce income taxes that would otherwise be payable currently; others increase income taxes that would otherwise be payable currently.

- **tinted or reflective glass or shading films**

Types of glass or a shading film applied to glass that, when installed on the exterior of a building, reduces the rates of solar penetration into the building. Includes Low E Glass.

- **tipping fee**

Price charged to deliver municipal solid waste to a landfill, waste-to-energy facility, or recycling facility.

- **tipple**

A central facility used in loading coal for transportation by rail or truck.

- **tolling arrangement**

Contract arrangement under which a raw material or intermediate product stream from one company is delivered to the production facility of another company in exchange for the equivalent volume of finished products and payment of a processing fee.

- **toluene (c6h5ch3)**
Colorless liquid of the aromatic group of petroleum hydrocarbons, made by the catalytic reforming of petroleum naphthas containing methyl cyclohexane. A high-octane gasoline-blending agent, solvent, and chemical intermediate, and a base for TNT (explosive).

- **ton mile**
The product of the distance that freight is hauled, measured in miles, and the weight of the cargo being hauled, measured in tons. Thus, moving one ton for one mile generates one ton mile.

- **topping cycle**
A boiler produces steam to power a turbine-generator to produce electricity. The steam leaving the turbine is used in thermal applications such as space heating and/or cooling or delivered to other end user(s).

- **total discoveries**
The sum of extensions, new reservoir discoveries in old fields, and new field discoveries, that occurred during the report year.

- **total gas in storage**
The sum of base gas and working gas.

- **total liquid hydrocarbon reserves**
The sum of crude oil and natural gas liquids reserves volumes.

- **total operated basis**
The total reserves or production associated with the wells operated by an individual operator. This is also commonly known as the "gross operated" or "8/8ths" basis.

- **toxic waste dumping at sea**
  - Large quantities of oil, nuclear waste, plastics and other debris, and a vast variety of chemical contaminants, are dumped or discharged at sea.
  - Many directly poison marine life or cause chronic disease, reproductive failure, or deformities.
  - The sources of marine pollution include commercial, military and recreational shipping and boating; run-off from urban streets and agricultural fields; oil drilling installations; and industries and sewage treatment plants.
  - The discharge of nutrients and other substances – for example, in human wastes in sewage; fertilizers and animal waste from farm runoff, and air emissions from coal- and oil-burning industries – is polluting coastal waters with excess nutrients.
  - These substances cause oxygen depletion of nearshore waters, harmful algal blooms, and dramatic reductions in the richness of sea-life communities.

Nutrient pollution

Pollution of the seas by nutrients is a serious problem. Dissolved inorganic nitrates and phosphates from municipal sewage plants, agricultural runoff and algae. The algae in turn can produce toxins that affect fish or shellfish, and reduce the oxygen content of water, stressing marine organisms.

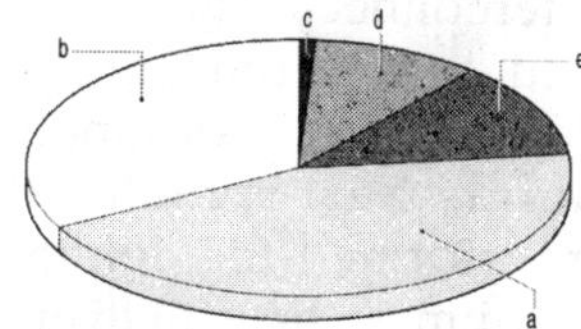

Sources of marine pollution
a nutrient pollution and land-derived pollutants, including industrial effluent and sewage-42 percent
b air-based pollution, including sulfur and nitrogen oxides returned in rain - 35 percent
c pollution from offshore production, including oil and gas extraction and mining - 1 percent
d deliberate dumping at sea of garbage and industrial waste - 10 percent
e transportation-garbage, including deliberate and accidental discharge of oil - 12 percent

■ toxic waste dumping on land

- Europe produces more than 2.5 billion tons of solid waste a year, and every day the inhabitants of New York City throw away approximately 26,000 tons.
- Land pollution occurs because of:
  1. Deposition of solid waste
  2. Accumulation of non-biodegradable materials
  3. Toxification of chemicals into poisons
  4. Alteration of soil chemical composition
- In most countries, most garbage is still buried in landfill sites.
- Landfill sites are a possible source of toxic chemicals and produce large quantities of methane gas.
- They must be managed so that pollutants do not seep into groundwater and should therefore be kept dry, but this slows down the rate of decomposition.
- Tyres constitute another problem, as they are virtually non-degradable and spread noxious fumes when burnt.
- Western Europe, the United States and Japan combined produce around 580 million tyres a year.

Sources and methods of land pollution

Sources	Methods
agriculture	build-up ot animal manures excessive input of chemical fertilizers dumping tainted crops on land illegally
mining and quarrying	blowing up mines with explosives using machinery which emits toxic by-products leakage to the ground
sewage sludge	leakage of sludge into the soil owing to improper sanitation system
dredged spoils	soil infertility caused by improper method of dredging at fertile land; the soil then becomes more prone to external pollution
household	accumulation of waste as a result of an improper system of disposal

Sources	Methods
	improper system of sanitation
demolition & construction	rubble or debris which is non-biodegradable undergoes chemical reactions, and increases soil toxicity if it is not settled in the soil
industrial	toxic or poisonous emissions of unfiltered or neutralized gases

Cross-section of typical landfill site

- The compacted garbage (between soil cover) will biodegrade naturally if it has been properly graded.
- The leachate pump contains a mixture of garbage liquid and rainwater which is pumped out and sent to a sewage treatment plant.

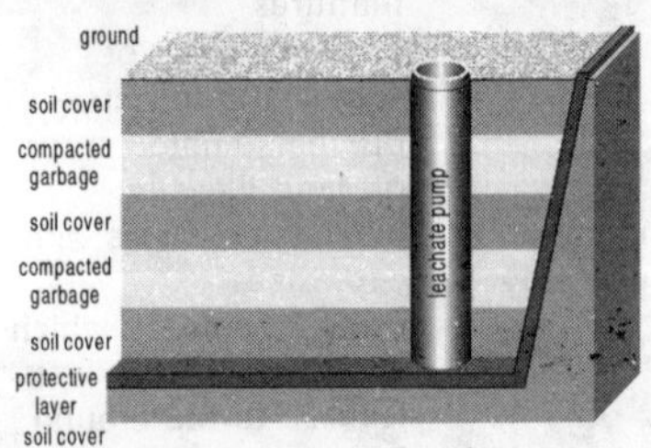

- **transfer capability**

The overall capacity of interregional or international power lines, together with the associated electrical system facilities, to transfer power and energy from one electrical system to another.

- **transfer price**

The monetary value assigned to products, services, or rights conveyed or exchanged between related parties, including those occurring between units of a consolidated entity.

- **transformer**

An electrical device for changing the voltage of alternating current.

- **transmission (electric) (verb)**

The movement or transfer of electric energy over an interconnected group of lines and associated equipment between points of supply and points at which it is transformed for delivery to consumers or is delivered to other electric systems. Transmission is considered to end when the energy is transformed for distribution to the consumer.

- **transmission and distribution loss**

Electric energy lost due to the transmission and distribution of electricity. Much of the loss is thermal in nature.

- **transmission circuit**

A conductor used to transport electricity from generating stations to load.

- **transmission line**

A set of conductors, insulators, supporting structures, and associated equipment used to move large quantities of power at high voltage, usually over long distances between a generating or receiving point and major substations or delivery points.

- **transmission network**
A system of transmission or distribution lines so cross-connected and operated as to permit multiple power supply to any principal point.

- **transmission system (electric)**
An interconnected group of electric transmission lines and associated equipment for moving or transferring electric energy in bulk between points of supply and points at which it is transformed for delivery over the distribution system lines to consumers or is delivered to other electric systems.

- **transmission type (engine)**
The transmission is the part of a vehicle that transmits motive force from the engine to the wheels, usually by means of gears for different speeds using either a hydraulic "torque-converter" (automatic) or clutch assembly (manual). On front-wheel drive cars, the transmission is often called a "transaxle." Fuel efficiency is usually higher with manual rather than automatic transmissions, although modern, computer-controlled automatic transmissions can be efficient.

- **transmitting utility**
A regulated entity which owns and may construct and maintain wires used to transmit wholesale power. It may or may not handle the power dispatch and coordination functions. It is regulated to provide non-discriminatory connections, comparable service, and cost recovery. According to the Energy Policy Act of 1992, it includes any electric utility, qualifying cogeneration facility, qualifying small power production facility, or Federal power marketing agency which owns or operates electric power transmission facilities which are used for the sale of electric energy at wholesale.

- **transport**
Movement of natural, synthetic, and/or supplemental gas between points beyond the immediate vicinity of the field or plant from which produced except (1) for movements through well or field lines to a central point for delivery to a pipeline or processing plant within the same state or (2) movements from a citygate point of receipt to consumers through distribution mains.

- **transportation agreement**
Any contractual agreement for the transportation of natural and/or supplemental gas between points for a fee.

- **transportation sector**
An energy-consuming sector that consists of all vehicles whose

primary purpose is transporting people and/or goods from one physical location to another. included are automobiles; trucks; buses; motorcycles; trains, subways, and other rail vehicles; aircraft; and ships, barges, and other waterborne vehicles. vehicles whose primary purpose is not transportation (e.g., construction cranes and bulldozers, farming vehicles, and warehouse tractors and forklifts) are classified in the sector of their primary use.

- **transported gas**

Natural gas physically delivered to a building by a local utility, but not purchased from that utility. A separate transaction is made to purchase the volume of gas, and the utility is paid for the use of its pipeline to deliver the gas. Also called "Direct-Purchase Gas," "Spot Market Gas," "Spot Gas," "Gas for the Account of Others", and "Self-Help Gas."

- **transporter**

The party or parties, other than buyer or seller, owning the facilities by which gas or LNG is physically transferred between buyer and seller.

- **transshipment**

A method of ocean transportation whereby ships off-load their oil cargo to a deepwater terminal, floating storage facility, temporary storage, or to one or more smaller tankers from which or in which the oil is then transported to a market destination.

- **treating plant**

A plant designed primarily to remove undesirable impurities from natural gas to render the gas marketable.

- **trillion btu**

Equivalent to 1,000,000,000,000 or 10 to the 12th power Btu.

- **troposphere**

The inner layer of the atmosphere below about 15 kilometers, within which there is normally a steady decrease of temperature with increasing altitude. Nearly all clouds form and weather conditions manifest themselves within this region. Its thermal structure is caused primarily by the heating of the earth's surface by solar radiation, followed by heat transfer through turbulent mixing and convection.

- **trough**

High-temperature (180+) concentrator with one axis-tracking.

- **trunk line**

A main pipeline.

- **turbine**

A machine for generating rotary mechanical power from the energy of a stream of fluid (such as water, steam, or hot gas).

Turbines convert the kinetic energy of fluids to mechanical energy through the principles of impulse and reaction, or a mixture of the two.

- **type of drive (vehicle)**

Refers to which wheels the engine power is delivered to, the so-called "drive wheels." Rear-wheel drive has drive wheels on the rear of the vehicle. Front-wheel drive, a newer technology, has drive wheels on the front of the vehicle. Four-wheel drive uses all four wheels as drive wheels and is found mostly on Jeep-like vehicles and trucks, though it is becoming increasingly more common on station wagons and vans.

- **ultimate analysis**

Determines the amount of carbon, hydrogen, oxygen, nitrogen, and sulfur. heating value is determined in terms of btu, both on an as received basis (including moisture) and on a dry basis.

- **ultimate customer**

A customer that purchases electricity for its own use and not for resale.

- **ultraviolet**

Electromagnetic radiation in the wavelength range of 4 to 400 nanometers.

- **unaccounted for (crude oil)**

Represents the arithmetic difference between the calculated supply and the calculated disposition of crude oil. The calculated supply is the sum of crude oil production plus imports minus changes in crude oil stocks. The calculated disposition of crude oil is the sum of crude oil input to refineries, crude oil exports, crude oil burned as fuel, and crude oil losses.

- **unaccounted for (natural gas)**

Represents differences between the sum of the components of natural gas supply and the sum of components of natural gas disposition. These differences may be due to quantities lost or to the effects of data reporting problems. Reporting problems include differences due to the net result of conversions of flow data metered at varying temperatures and pressure bases and converted to a standard temperature and pressure base; the effect of variations in company accounting and billing practices; differences between billing cycle and calendar-period time frames; and imbalances resulting from the merger of data reporting systems that vary in scope, format, definitions, and type of respondents.

- **unbundling**

Separating vertically integrated monopoly functions into their component parts for the purpose of separate service offerings.

- **uncompleted wells, equipment, and facilities costs**

The costs incurred to (1) drill and equip wells that are not yet completed, and (2) acquire or construct equipment and facilities that are not yet completed and installed.

- **unconsolidated entity**

A firm directly or indirectly controlled by a parent but not consolidated with the parent for purposes of financial statements prepared in accordance with generally accepted accounting principles. An unconsolidated entity includes any firm consolidated with the unconsolidated entity for purposes of financial statements prepared in accordance with generally accepted accounting principles historically and consistently applied. An individual shall be deemed to control a firm that is directly or indirectly controlled by him or by his father, mother, spouse, children, or grandchildren.

- **underground gas storage reservior capacity**

Interstate company reservoir capacities are those certificated by the Federal Energy Regulatory Commission. Independent producer and intrastate company reservior capacities are reported as developed capacity.

- **underground gas storage**

The use of sub-surface facilities for storing gas that has been transferred from its original location. The facilities are usually hollowed-out salt domes, geological reservoirs (depleted oil or gas fields) or water-bearing sands topped by an impermeable cap rock (aquifer).

- **underground mine**

A mine where coal is produced by tunneling into the earth to the coal bed, which is then mined with underground mining equipment such as cutting machines and continuous, long wall, and short wall mining machines. Underground mines are classified according to the type of opening used to reach the coal, i.e., drift (level tunnel), slope (inclined tunnel), or shaft (vertical tunnel).

- **underground mining equipment**

A **coal-cutting machine** is used in conventional mining to undercut, top cut, or shear the coal face so that coal can be fractured easily when blasted. It cuts 9 to 13 feet into the bed.

Continuous auger machine is used in mining coal beds less than 3 feet thick. The auger has a cutting depth of about 5 feet and is 20 to 28 inches in diameter.

Continuous auger mining usually uses a conveyor belt to haul the coal to the surface.

Continuous mining machine, used during continuous mining, cuts or rips coal from the face and loads it into shuttle cars or conveyors in one operation. It eliminates the use of blasting devices and performs many functions of other equipment such as drills, cutting machines, and loaders. A continuous mining machine typically has a turning "drum" with sharp bits that cut and dig out the coal for 16 to 22 feet before mining stops so that the mined area can be supported with roof bolts. This machine can mine coal at the rate of 8 to 15 tons per minute. There are of two types of **conveyor systems**: 1. A **mainline conveyor**, which is usually a permanent installation that carries coal to the surface. 2. A **section conveyor**, which connects the working face to the mainline conveyor. **Face drill** is used in conventional mining to drill shotholes in the coalbed for explosive charges. **Loading machine** is used in conventional mining to scoop broken coal from the working area and load it into a shuttle car, which hauls the coal to mine cars or conveyors for delivery to the surface. **Longwall mining machine** shears coal from a long straight coal face (up to about 700 feet) by working back and forth across the face under a movable, hydraulic-jack roof-support system. The broken coal is transported by converyor. Longwall machines can mine coal at the rate of 1,000 tons per shift. **Mine locomotive**, operating on tracks, is used to haul mine cars containing coal and other material, and to move personnel in specially designed "mantrip" cars. Large locomotives can haul more than 20 tons at a speed of about 10 miles per hour. Most mine locomotives run on electricity provided by a trolley wire; some are battery-powered. **Ram car or shuttle ram** is a rubber-tired haulage vehicle that is unloaded through the use of a movable steel plate located at the back of the haulage bed. **Roof-bolting machine, or roof bolter**, is used to drill holes and place bolts to support the mine roof. Roof bolting units can be installed on a continuous mining machine. **Scoop** is a rubber-tired haulage vehicle used in thin coalbeds. **Short wall mining machine** generally is a continuous-mining machine used with a powered, self-advancing roof support system. It shears coal from a short coal face (up to about 150 feet long). The broken coal is hauled by shuttle cars to a conveyor belt. **Shuttle car** is a rubber-tired haulage vehicle that is unloaded by a built-in conveyor.

▪ underground mining methods

A **drift mine** is driven horizontally into coal that is exposed or accessible in a hillside. In a hydraulic mine, high-pressure water jets break the coal from a steeply inclined, thick coalbed that would be difficult to mine with the usual underground methods. The coal is then transported to the surface by a system of flumes or by pipeline. Although currently not in commercial use in the United States, hydraulic mining is used in western Canada.

A **punch mine** is a type of small drift mine used to recover coal from strip-mine high walls or from small, otherwise uneconomical, coal deposits. A shaft mine is driven vertically to the coal deposit. A slope mine is driven at an angle to reach the coal deposit.

In a **room-and-pillar mining** system, the most common method, the mine roof, is supported mainly by coal pillars left at regular intervals. Rooms are places where the coal is mined; pillars are areas of coal left between the rooms. Room-and-pillar mining is done either by 1) conventional mining, which involves a series of operations that require cutting the working face of the coal bed so that it breaks easily when blasted with explosives or high-pressure air, and then loading the broken coal or 2) continuous mining, in which a continuous mining machine extracts and removes coal from the working face in one operation. When a section of a mine has been fully developed, additional coal is extracted by mining the supportive pillars until the roof caves in; the procedure is called room-and-pillar retreat mining.

In a **long wall mining** system, long sections of coal, up to about 700 feet, are removed and no pillars are left to support the mined-out areas. The working area is protected by a movable, powered roof support system. The caved area (gob) compacts and, after initial subsidence, supports the overlying strata. Long wall mining is used where the coal bed is thick and generally flat, where surface subsidence is acceptable.

A **short wall mining** system generally refers to the room-and-pillar mining in which the working face is wider than usual but smaller (less than 150 feet) than that in long wall mining.

Roof support and mine ventilation are paramount in all underground mining operations. Roof bolting is the principal method of supporting the mine roof. In roof bolting, long bolts, 2 to 10 feet long with an expansion shell or with resin grouting are placed in the mine roof. The bolts reinforce the roof by pulling together rock strata to make a strong beam or by fastening weak strata to strong strata. Mine ventilation, accomplished with fans, is essential to supply fresh air and to remove gases and dust from the mine. To reduce the possibility of coal dust explosions, rock dust is sprayed in an underground coal mine. Rock dust is a very fine noncombustible material (pulverized limestone).

- **underground storage injections**
  Gas from extraneous sources put into underground storage reservoirs.

- **underground storage**
  The storage of natural gas in underground reservoirs at a different location from which it was produced.

- **underground storage withdrawals**
  Gas removed from underground storage reservoirs.

- **undifferentiated/unspecified reserves and production**
  Reserves and production that are not separable by FERC production areas or by states.
  Undifferentiated and unspecified reserves consist only of company-owned gas in underground storage.

- **undiscovered recoverable reserves (crude oil and natural gas)**
  Those economic resources of crude oil and natural gas, yet undiscovered, that are estimated to exist in favorable geologic settings.

- **undiscovered resources (coal)**
  Unspecified bodies of coal surmised to exist on the basis of broad geologic knowledge and theory. Undiscovered resources include beds of bituminous coal and anthracite 14 inches or more thick and beds of subbituminous coal and lignite 30 inches or more thick that are presumed to occur in unmapped and unexplored areas to depths of 6,000 feet. The speculative and hypothetical resource categories comprise undiscovered resources.

- **unfilled requirements**
  Requirements not covered by usage of inventory or supply contracts in existence as of January 1 of the survey year.

- **unfinished oils**
  All oils requiring further processing, except those requiring only mechanical blending. Unfinished oils are produced by partial refining of crude oil and include naphthas and lighter oils, kerosene and light gas oils, heavy gas oils, and residuum.

- **unfractionated streams]**
  Mixtures of unsegregated natural gas liquid components, excluding those in plant condensate. This product is extracted from natural gas.

- **unglazed solar collector**
  A solar thermal collector that has an absorber that does not have a glazed covering. Solar swimming

pool heater systems usually use unglazed collectors because they circulate relatively large volumes of water through the collector and capture nearly 80 percent of the solar energy available.

- **uniform system of accounts**

Prescribed financial rules and regulations established by the Federal Energy Regulatory Commission for utilities subject to its jurisdiction under the authority granted by the Federal Power Act.

- **unit price**

Total revenue derived from the sale of product during the reference month divided by the total volume sold; also known as the weighted average price. Total revenue should exclude all taxes but include transportation costs that were paid as part of the purchase price.

- **unit value, consumption**

Total price per specified unit, including all taxes, at the point of consumption.

- **unit value, wellhead**

The wellhead sales price, including charges for natural gas plant liquids subsequently removed from the gas; gathering and compression charges; and state production, severance, and/or similar charges.

- **unleaded gasoline**

Contains not more than 0.05 gram of lead per gallon and not more than 0.005 gram of phosphorus per gallon. Premium regular and intermediate grades are included, depending on the octane rating.

- **unprocessed gas**

Natural gas that has not gone through a processing plant.

- **unregulated entity**

For the purpose of EIA's data collection efforts, entities that do not have a designated franchised service area and that do not file forms listed in the Code of Federal Regulations, Title 18, Part 141, are considered unregulated entities. This includes qualifying cogenerators, qualifying small power producers, and other generators that are not subject to rate regulation, such as independent power producers.

- **unscheduled outage service**

Power received by a system from another system to replace power from a generating unit forced out of service.

- **uranium (u)**

A heavy, naturally radioactive, metallic element (atomic number 92). Its two principally occurring isotopes are uranium-235 and uranium-238. Uranium-235 is indispensable to the nuclear industry because it is the only isotope existing in nature, to any appreciable extent, that is fissionable by thermal neutrons. Uranium-238 is also important because it absorbs neutrons to produce a radioactive isotope that subsequently decays to the isotope plutonium-239, which also is fissionable by thermal neutrons.

- **uranium concentrate**

A yellow or brown powder obtained by the milling of uranium ore, processing of in situ leach mining solutions, or as a byproduct of phosphoric acid production.

- **uranium deposit**

A discrete concentration of uranium mineralization that is of possible economic interest.

- **uranium endowment**

The uranium that is estimated to occur in rock with a grade of at least 0.01 percent U_3O_8. The estimate of the uranium endowment is made before consideration of economic availability of any associated uranium resources.

- **uranium hexaflouride (uf6)**

A white solid obtained by chemical treatment of U_3O_8 and which forms a vapor at temperatures above 56 degrees Centigrade. UF_6 is the form of uranium required for the enrichment process.

- **uranium importation**

The actual physical movement of uranium from a location outside the United States to a location inside the United States.

- **uranium mill**

A plant where uranium is separated from ore taken from mines.

- **uranium mill tailings radiation control act (umtra) of 1978**

The act that directed the Department of Energy to provide for stabilization and control of the uranium mill tailings from inactive sites in a safe and environmentally sound manner to minimize radiation health hazards to the public. It authorized the Department to undertake remedial actions at 24 designated inactive uranium-processing sites and at an estimated 5,048 vicinity properties.

- **uranium mill tailings**

The sand-like materials left over from the separation of uranium from its ore. More than 99 percent of the ore becomes tailings.

- **uranium ore**

Rock containing uranium mineralization in concentrations that can be mined economically, typically one to four pounds of U_3O_8 per ton or 0.05 percent to 0.2 percent U_3O_8.

- **uranium oxide**

Uranium concentrate or yellowcake. Abbreviated as U_3O_8.

■ **uranium property**

A specific piece of land with uranium reserves that is held for the ultimate purpose of economically recovering the uranium. The land can be developed for production or undeveloped.

■ **uranium reserves**

Estimated quantities of uranium in known mineral deposits of such size, grade, and configuration that the uranium could be recovered at or below a specified production cost with currently proven mining and processing technology and under current law and regulations. Reserves are based on direct radiometric and chemical measurements of drill holes and other types of sampling of the deposits. Mineral grades and thickness, spatial relationships, depths below the surface, mining and reclamation methods, distances to milling facilities, and amenability of ores to processing are considered in the evaluation. The amount of uranium in ore that could be exploited within the chosen forward-cost levels are estimated in accordance with conventional engineering practices.

■ **uranium resource categories (international)**

Three categories of uranium resources defined by the international community to reflect differing levels of confidence in the existence of the resources. Reasonably assured resources (RAR), estimated additional resources (EAR), and speculative resources (SR) are described below.

Reasonably assured resources (RAR) Uranium that occurs in known mineral deposits of such size, grade, and configuration that it could be recovered within the given production cost ranges, with currently proven mining and processing technology. Estimates of tonnage and grade are based on specific sample data and measurements of the deposits and on knowledge of deposit characteristics. Note: **RAR** corresponds to DOE's uranium reserves category.

Estimated additional resources (EAR) Uranium in addition to **RAR** that is expected to occur, mostly on the basis of geological evidence, in extensions of well-explored deposits, in little-explored deposits, and in undiscovered deposits believed to exist along well-defined geological trends with known deposits. This uranium can subsequently be recovered within the given cost ranges. Estimates of tonnage and grade are based on available sampling data and on knowledge of the deposit characteristics, as determined in the best-known parts of the deposit or in similar deposits.

Speculative resources (SR)
Uranium in addition to **EAR** that is thought to exist, mostly on the basis of indirect evidence and geological extrapolations, in deposits discoverable with existing exploration techniques. The location of deposits in this category can generally be specified only as being somewhere within given regions or geological trends. The estimates in this category are less reliable than estimates of **RAR** and **EAR**.

- **usage agreement**

Contracts held by enrichment customers that allow feed material to be stored at the enrichment plant site in advance of need.

- **used and useful**

A concept used by regulators to determine whether an asset should be included in the utility's rate base. This concept requires that an asset currently provide or be capable of providing a needed service to customers.

- **useful thermal output**

The thermal energy made available in a combined-heat-and-power system for use in any industrial or commercial process, heating or cooling application, or delivered to other end users, i.e., total thermal energy made available for processes and applications other than electrical generation.

- **utility demand-side management costs**

The costs incurred by the utility to achieve the capacity and energy savings from the Demand-Side Management (DSM) Program. Costs incurred by consumers or third parties are to be excluded. The costs are to be reported in nominal dollars in the year in which they are incurred, regardless of when the savings occur. The utility costs are all the annual expenses (labor, administrative, equipment, incentives, marketing, monitoring and evaluation, and other) incurred by the utility for operation of the DSM Program, regardless of whether the costs are expensed or capitalized. Lump-sum capital costs (typically accrued over several years prior to start up) are not to be reported. Program costs associated with strategic load growth activities are also to be excluded.

- **utility distribution companies**

The entities that will continue to provide regulated services for the distribution of electricity to customers and serve customers who do not choose direct access. Regardless of where a consumer chooses to purchase power, the customer's current utility, also known as the utility distribution company, will deliver the power to the consumer.

- **utility generation**
Generation by electric systems engaged in selling electric energy to the public.

- **utility-sponsored conservation program**
Any program sponsored by an electric and/or natural gas utility to review equipment and construction features in buildings and advise on ways to increase the energy efficiency of buildings. Also included are utility-sponsored programs to encourage the use of more energy-efficient equipment. Included are programs to improve the energy efficiency in the lighting system or building equipment or the thermal efficiency of the building shell.

- **vacuum distillation**
Distillation under reduced pressure (less the atmospheric), which lowers the boiling temperature of the liquid being distilled. This technique with its relatively low temperatures prevents cracking or decomposition of the charge stock.

- **value (of shipments)**
The value received for the complete systems at the company's net billing price, freight-on-board factory, including charges for cooperative advertising and warranties. This does not include excise taxes, freight or transportation charges, or installation charges.

- **value added by manufacture**
A measure of manufacturing activity that is derived by subtracting the cost of materials (which covers materials, supplies, containers, fuel, purchased electricity, and contract work) from the value of shipments. This difference is then adjusted by the net change in finished goods and work-in-progress between the beginning- and end-of-year inventories.

- **vapor displacement**
The release of vapors that had previously occupied space above liquid fuels stored in tanks. These releases occur when tanks are emptied and filled.

- **vapor retarder**
A material that retards the movement of water vapor through a building element (walls, ceilings) and prevents insulation and structural wood from becoming damp and metals from corroding. Often applied to insulation batts or separately in the form of treated papers, plastic sheets, and metallic foils.

- **vapor-dominated geothermal system**
A conceptual model of a hydrothermal system where steam pervades the rock and is the pressure-controlling fluid phase.

- **variable air volume (vav) system on the heating and cooling system**

A means of varying the amount of conditioned air to a space. A variable air volume system maintains the air flow at a constant temperature, but supplies varying quantities of conditioned air in different parts of the building according to the heating and cooling needs.

- **variable-speed wind turbines**

Turbines in which the rotor speed increases and decreases with changing wind speed, producing electricity with a variable frequency.

- **vehicle fuel consumption**

Vehicle fuel consumption is computed as the vehicle miles traveled divided by the fuel efficiency reported in miles per gallon (MPG). Vehicle fuel consumption is derived from the actual vehicle mileage collected and the assigned MPGs obtained from EPA certification files adjusted for on-road driving. The quantity of fuel used by vehicles.

- **vehicle fuel expenditures**

The cost, including taxes, of the gasoline, gasohol, or diesel fuel added to the vehicle's tank. Expenditures do not include the cost of oil or other items that may have been purchased at the same time as the vehicle fuel.

- **vehicle identification number (vin)**

A set of codes, usually alphanumeric characters, assigned to a vehicle at the factory and inscribed on the vehicle. When decoded, the VIN provides vehicle characteristics. The VIN is used to help match vehicles to the EPA certification file for calculating MPGs.

- **vehicle importer**

An original vehicle manufacturer (of foreign or domestic ownership) that imports vehicles as finished products into the United States.

- **vehicle miles traveled (vmt)**

The number of miles traveled nationally by vehicles for a period of 1 year. VMT is either calculated using two odometer readings or, for vehicles with less than two odometer readings, imputed using a regression estimate.

- **vented**

Gas released into the air on the production site or at processing plants.

- **vented/flared**

Gas that is disposed of by releasing (venting) or burning (flaring).

- **ventilation system**

A method for reducing methane concentrations in coal mines to non-explosive levels by blowing air across the mine face and using large exhaust fans to remove methane while mining operations proceed.

- **vertical integration**

The combination within a firm or business enterprise of one or more stages of production or distribution. In the electric industry, it refers to the historical arrangement whereby a utility owns its own generating plants, transmission system, and distribution lines to provide all aspects of electric service.

- **vertical-axis wind turbine (vawt)**

A type of wind turbine in which the axis of rotation is perpendicular to the wind stream and the ground.

- **vessel**

A ship used to transport crude oil, petroleum products, or natural gas products. Vessel categories are as follows: Ultra Large Crude Carrier (ULCC), Very Large Crude Carrier (VLCC), Other Tanker, and Specialty Ship (LPG/LNG).

- **vessel bunkering**

Includes sales for the fueling of commercial or private boats, such as pleasure craft, fishing boats, tugboats, and ocean-going vessels, including vessels operated by oil companies.

- **vin (vehicle identification number)**

A set of about 17 codes, combining letters and numbers, assigned to a vehicle at the factory and inscribed on a small metal label attached to the dashboard and visible through the windshield. The VIN is a unique identifier for the vehicle and therefore is often found on insurance cards, vehicle registrations, vehicle titles, safety or emission certificates, insurance policies, and bills of sale. The coded information in the VIN describes characteristics of the vehicle such as engine size and weight.

- **virgin coal**

Coal that has not been accessed by mining.

- **visbreaking**

A thermal cracking process in which heavy atmospheric or vacuum-still bottoms are cracked at moderate temperatures to increase production of distillate products and reduce viscosity of the distillation residues.

- **volatile matter**

Those products, exclusive of moisture, given off by a material as gas or vapor. Volatile matter is determined by heating the coal to 950 degrees Centigrade under carefully controlled conditions and measuring the weight loss, excluding weight of moisture driven off at 105 degrees Centigrade.

- **volatile organic compounds**

Organic compounds that participate in atmospheric photochemical reactions.

■ **volatile solids**

A solid material that is readily decomposable at relatively low temperatures.

■ **volt (v)**

The volt is the International System of Units (SI) measure of electric potential or electromotive force. A potential of one volt appears across a resistance of one ohm when a current of one ampere flows through that resistance. Reduced to SI base units, $1\ V = 1$ kg times m^2 times s^{-3} times A^{-1} (kilogram meter squared per second cubed per ampere).

■ **voltage reduction**

Any intentional reduction of system voltage by 3 percent or greater for reasons of maintaining the continuity of service of the bulk electric power supply system.

■ **voltage**

The difference in electrical potential between any two conductors or between a conductor and ground. It is a measure of the electric energy per electron that electrons can acquire and/or give up as they move between the two conductors.

■ **wafer**

A thin sheet of semiconductor (photovoltaic material) made by cutting it from a single crystal or ingot.

■ **walk-in refrigeration units**

Refrigeration/freezer units within a building that are large enough to walk into. They may be portable or permanent, such as a meat storage locker in a butcher store. Walk-in units may or may not have a door, plastic strips, or other flexible covers.

■ **wall insulation**

Insulating materials within or on the walls between heated areas of the building and unheated areas or the outside. The walls may separate air-conditioned areas from areas not air-conditioned.

■ **warranty contracts**

Gas purchase agreements for the sale of natural gas by a producer to a pipeline company wherein the producer warrants it will have available sufficient gas supplies to meet its commitments over the life of the contract. Generally, the producer does not dedicate gas reserves underlying any specific acreage, lease, or fields to the agreement. Substitution of various sources of gas supply may be permitted according to the terms of the contract. Warranty contracts, by their terms, may vary from the above.

■ **waste coal**

Usable coal material that is a byproduct of previous processing operations or is recaptured from what would otherwise be refuse. Examples include anthracite culm, bituminous gob, fine coal, lignite waste, coal recovered from a refuse bank or slurry dam, and coal recovered by dredging.

- **waste energy**
  Municipal solid waste, landfill gas, methane, digester gas, liquid acetonitrile waste, tall oil, waste alcohol, medical waste, paper pellets, sludge waste, solid byproducts, tires, agricultural byproducts, closed loop biomass, fish oil, and straw used as fuel.

- **waste heat boiler**
  A boiler that receives all or a substantial portion of its energy input from the combustible exhaust gases from a separate fuel-burning process.

- **waste heat recovery**
  Any conservation system whereby some space heating or water heating is done by actively capturing byproduct heat that would otherwise be ejected into the environment. In commercial buildings, sources of water- heat recovery include refrigeration/ air-conditioner compressors, manufacturing or other processes, data processing centers, lighting fixtures, ventilation exhaust air, and the occupants themselves. Not to be considered is the passive use of radiant heat from lighting, workers, motors, ovens, etc., when there are no special systems for collecting and redistributing heat.

- **waste materials**
  Otherwise discarded combustible materials that, when burned, produce energy for such purposes as space heating and electric power generation. The size of the waste may be reduced by shredders, grinders, or hammer mills. Noncombustible materials, if any, may be removed. The waste may be dried and then burned, either alone or in combination with fossil fuels.

- **waste oils and tar**
  Petroleum-based materials that are worthless for any purpose other than fuel use.

- **wastewater, domestic and commercial**
  Wastewater (sewage) produced by domestic and commercial establishments.

- **wastewater, industrial**
  Wastewater produced by industrial processes.

- **water bed heater**
  An appliance that uses an electric resistance coil to maintain the temperature of the water in a water bed at a comfortable level.

- **water conditions**
  The status of the water supply and associated water in pondage and reservoirs at hydroelectric plants.

water heated in furnace

Some furnaces provide hot water as well as heat the home. The water is heated by a coil that is part of the furnace. There is no separate hot water tank.

water heater

An automatically controlled, thermally insulated vessel designed for heating water and storing heated water at temperatures less than 180 degrees Fahrenheit.

water heating dsm programs

These are demand-side management (DSM) programs designed to promote increased efficiency in water heating, including water heater insulation wraps.

water heating equipment

Automatically controlled, thermal insulated equipment designed for heating and storing heated water at temperatures less than 180 degrees Fahrenheit for other than space heating purposes.

water pollution abatement equipment

Equipment used to reduce or eliminate waterborne pollutants, including chlorine, phosphates, acids, bases, hydrocarbons, sewage, and other pollutants. Examples of water pollution abatement structures and equipment include those used to treat thermal pollution; cooling, boiler, and cooling tower blow down water; coal pile runoff; and fly ash waste water. Water pollution abatement excludes expenditures for treatment of water prior to use at the plant.

water pumping

Photovoltaic modules/cells used for pumping water for agricultural, land reclamation, commercial, and other similar applications where water pumping is the main use.

water reservoir

A large inland body of water collected and stored above ground in a natural or artificial formation.

water source heat pump

A type of (geothermal) heat pump that uses well (ground) or surface water as a heat source. Water has a more stable seasonal temperature than air thus making for a more efficient heat source.

water turbine

A turbine that uses water pressure to rotate its blades; the primary types are the Pelton wheel, for high heads (pressure); the Francis turbine, for low to medium heads; and the Kaplan for a wide range of heads. Primarily used to power an electric generator.

water vapor

Water in a vaporous form, especially when below boiling temperature and diffused (e.g., in the atmosphere).

- **water well**

A well drilled to (1) obtain a water supply to support drilling or plant operations, or (2) obtain a water supply to be used in connection with an improved recovery program.

- **water wheel**

A wheel that is designed to use the weight and/or force of moving water to turn it, primarily to operate machinery or grind grain.

- **waterway**

A river, channel, canal, or other navigable body of water used for travel or transport.

- **watt (w)**

The unit of electrical power equal to one ampere under a pressure of one volt. A Watt is equal to 1/746 horsepower.

- **watt hour (wh)**

The electrical energy unit of measure equal to one watt of power supplied to, or taken from, an electric circuit steadily for one hour.

- **wattmeter**

A device for measuring power consumption.

- **wave energy**

Kinetic energy (movement) exists in the moving waves of the ocean. That energy can be used to power a turbine. In this simple example, to the right, the wave rises into a chamber. The rising water forces the air out of the chamber. The moving air spins a turbine which can turn a generator.

When the wave goes down, air flows through the turbine and back into the chamber through doors that are normally closed.

This is only one type of wave-energy system. Others actually use the up and down motion of the wave to power a piston that moves up and down inside a cylinder. That piston can also turn a generator.

Most wave-energy systems are very small. But, they can be used to power a warning buoy or a small light house.

- **wax**

A solid or semi-solid material derived from petroleum distillates or residues by such treatments as chilling, precipitating with a solvent, or de-oiling. It is a light-colored, more-or-less translucent crystalline mass, slightly greasy to the touch, consisting of a mixture of solid hydrocarbons in which the paraffin series predominates. Includes all marketable wax, whether crude scale or fully refined. The three grades included are microcrystalline, crystalline-fully refined, and crystalline-other. The conversion factor is 280 pounds per 42 U.S. gallons per barrel.

- **weather stripping or caulking**

Any of several kinds of crack-filling material around any windows or doors to the outside used to reduce the passage of air and moisture around moveable parts of a door or window. Weather stripping is available in strips or rolls of metal, vinyl, or foam rubber and can be applied on the inside or outside of a building.

- **weir**

A dam in a waterway over which water flows and that serves to raise the water level or to direct or regulate flow.

- **well**

A hole drilled in the earth for the purpose of (1) finding or producing **crude oil** or **natural gas**; or (2) producing services related to the production of crude or natural gas.

- **well water for cooling**

A means of cooling that uses water from a well drilled specifically for that purpose. The subterranean temperature of the water stays at a relatively constant temperature. Where water is abundant, it provides a means of getting 55-degree Fahrenheit water with no mechanical cooling. Used usually for heat rejection in a water source heat pump.

- **wellhead price**

The value at the mouth of the well. In general, the wellhead price is considered to be the sales price obtainable from a third party in an arm's length transaction. Posted prices, requested prices, or prices as defined by lease agreements, contracts, or tax regulations should be used where applicable.

- **wellhead**

The point at which the crude (and/or natural gas) exits the ground. Following historical precedent, the volume and price for crude oil production are labeled as "wellhead," even though the cost and volume are now generally measured at the lease boundary. In the context of domestic crude price data, the term "wellhead" is the generic term used to reference the production site or lease property.

- **wet bottom boiler**

Slag tanks are installed usually at the furnace throat to contain and remove molten ash.

- **wet natural gas**

A mixture of hydrocarbon compounds and small quantities of various nonhydrocarbons existing in the gaseous phase or in solution with crude oil in porous rock formations at reservoir conditions. The principal hydrocarbons normally contained in the mixture are methane, ethane, propane, butane, and pentane. Typical nonhydrocarbon gases that may be present in reservoir natural gas are water vapor, carbon dioxide, hydrogen sulfide, nitrogen and trace amounts of helium. Under reservoir conditions, natural gas and its associated liquefiable portions occur either in a single gaseous phase in the reservoir or in solution with crude oil and are not distinguishable at the time as separate substances.

- **wheeling charge**

An amount charged by one electrical system to transmit the energy of, and for, another system or systems.

- **wheeling service**

The movement of electricity from one system to another over transmission facilities of interconnecting systems. Wheeling service contracts can be established between two or more systems.

- **white spirit**

A highly refined distillate with a boiling point range of about 150 degrees to 200 degrees Centigrade. It is used as a paint solvent and for dry-cleaning purposes.

- **whole-house cooling fan**

A mechanical/electrical device used to pull air out of an interior space; usually located in the highest location of a building, in the ceiling, and venting to the attic or directly to the outside.

- **wholesale competition**

A system whereby a distributor of power would have the option to buy its power from a variety of power producers, and the power producers would be able to compete to sell their power to a variety of distribution companies.

- **wholesale electric power market**

The purchase and sale of electricity from generators to resellers (retailers), along with the ancillary services needed to maintain reliability and power quality at the transmission level.

- **wholesale power market**

The purchase and sale of electricity from generators to resellers (who sell to retail customers), along with the ancillary services needed to maintain reliability and power quality at the transmission level.

- **wholesale price**

The rack sales price charged for No. 2 heating oil; that is, the price charged customers who purchase No. 2 heating oil free-on-board at a supplier's terminal and provide their own transportation for the product.

- **wholesale sales**

Energy supplied to other electric utilities, cooperatives, municipals, and Federal and state electric agencies for resale to ultimate consumers.

- **wholesale transmission services**

The transmission of electric energy sold, or to be sold, in the wholesale electric power market.

- **wholesale wheeling**

An arrangement in which electricity is transmitted from a generator to a utility through the transmission facilities of an intervening system.

- **wind energy**

Wind can be used to do work. The kinetic energy of the wind can be changed into other forms of energy, either mechanical energy or electrical energy.

When a boat lifts a sail, it is using wind energy to push it through the water. This is one form of work.

Farmers have been using wind energy for many years to pump water from wells using windmills like the one on the right.

In Holland, windmills have been used for centuries to pump water from low-lying areas.

Wind is also used to turn large grinding stones to grind wheat or corn, just like a water wheel is turned by water power.

Today, the wind is also used to make electricity.

Blowing wind spins the blades on a wind turbine — just like a large toy pinwheel. This device is called a wind turbine and not a windmill. A windmill grinds or mills grain, or is used to pump water.

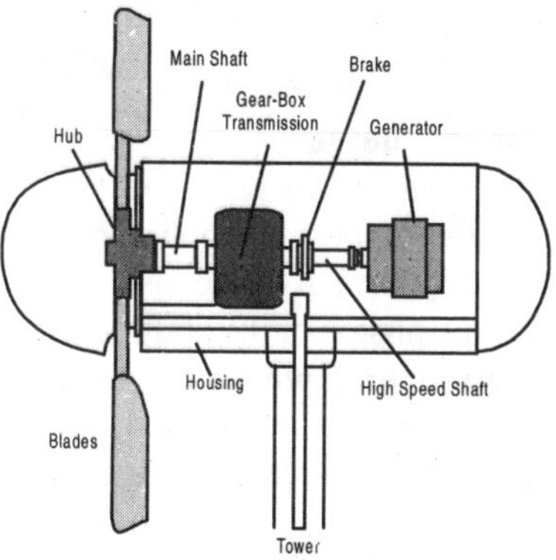

- **wind energy conversion system (wecs) or device**

An apparatus for converting the energy available in the wind to mechanical energy that can be used to power machinery (grain mills, water pumps) and to operate an electrical generator.

- **wind power plant**

A group of wind turbines interconnected to a common utility system through a system of transformers, distribution lines, and (usually) one substation. Operation, control, and maintenance functions are often centralized through a network of computerized monitoring systems, supplemented by visual inspection. This is a term commonly used in the United

States. In Europe, it is called a generating station.

- **wind turbine**

Wind energy conversion device that produces electricity; typically three blades rotating about a horizontal axis and positioned upwind of the supporting tower.

- **wires charge**

A broad term referring to fees levied on power suppliers or their customers for the use of the transmission or distribution wires.

wood and waste (as used at electric utilities) Wood energy, garbage, bagasse (sugarcane residue), sewerage gas, and other industrial, agricultural, and urban refuse used to generate electricity for distribution.

- **wood conversion to btu**

Converting cords of wood into a Btu equivalent is an imprecise procedure. The number of cords each household reports having burned is inexact, even with the more precise drawings provided, because the estimate requires the respondent to add up the use of wood over a 12-month period during which wood may have been added to the supply as well as removed. Besides errors of memory inherent in this task, the estimates are subject to problems in definition and perception of what a cord is. The nominal cord as delivered to a suburban residential buyer may differ from the dimensions of the standard cord. This difference is possible because wood is most often cut in lengths that are longer than what makes a third of a cord (16 inches) and shorter than what makes a half cord (24 inches). In other cases, wood is bought or cut in unusual units (for example, pickup-truck load, or trunk load). Finally, volume estimates are difficult to make when the wood is left in a pile instead of being stacked. Other factors that make it difficult to estimate the Btu value of the wood burned is that the amount of empty space between the stacked logs may vary from 12 to 40 percent of the volume. Moisture content may vary from 20 percent in dried wood to 50 percent in green wood. (Moisture reduces the useful Btu output because energy is used in driving off the moisture). Finally, some tree species contain twice the Btu content of species with the lowest Btu value. Generally, hard woods have greater Btu value than soft woods. Wood is converted to Btu at the rate of 20 million Btu per cord, which is a rough average that takes all these factors into account.

- **wood energy**

Wood and wood products used as fuel, including round wood (cord wood), limb wood, wood chips, bark, sawdust, forest residues, charcoal, pulp waste, and spent pulping liquor.

■ **wood pellets**

Sawdust compressed into uniform diameter pellets to be burned in a heating stove.

■ **working (top storage) gas**

The volume of gas in the reservoir that is in addition to the cushion or base gas. It may or may not be completely withdrawn during any particular withdrawal season. Conditions permitting, the total working capacity could be used more than once during any season.

■ **working interest**

An interest in a mineral property that entitles the owner of that interest to all of share of the mineral production from the property, usually subject to a royalty.

A working interest permits the owner to explore, develop, and operate the property. The working-interest owner bears the costs of exploration, development, and operation of the property and, in return, is entitled to a share of the mineral production from the property or to a share of the proceeds there from. It may be assigned to another party in whole or in part, or it may be divided into other special property interests. **Gross working interest.** The reporting company's working interest plus the proportionate share of any basic royalty interest or overriding royalty interest related to the working interest. **Net working interest.** The reporting company's working interest is not including any basic royalty or overriding royalty interests.

■ **working storage capacity**

The difference in volume between the maximum safe fill capacity and the quantity below which pump suction is ineffective (bottoms).

■ **world energy consumption**

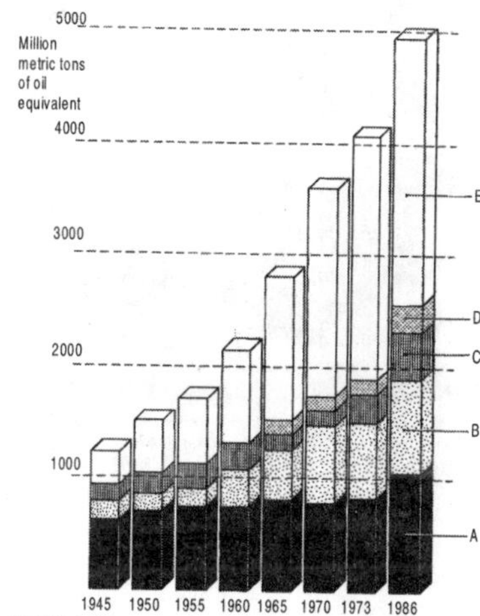

A Solid fuel • B Natural gas • C Hydroelectricity • D Nuclear electricity • E Oil

■ **world energy production**

- Global energy use has increased nearly 70 percent since 1971, and is projected to increase at more than two percent annually for the next 15 years.
- This increase will raise greenhouse gas emissions about 50 percent higher than

current levels, unless a concerted effort takes place to increase energy efficiency and move away from today's heavy reliance on fossil fuels.

- Fossil fuels account for some 80 percent of the world's primary energy consumption, and supply roughly 90 percent of the world's commercial energy.

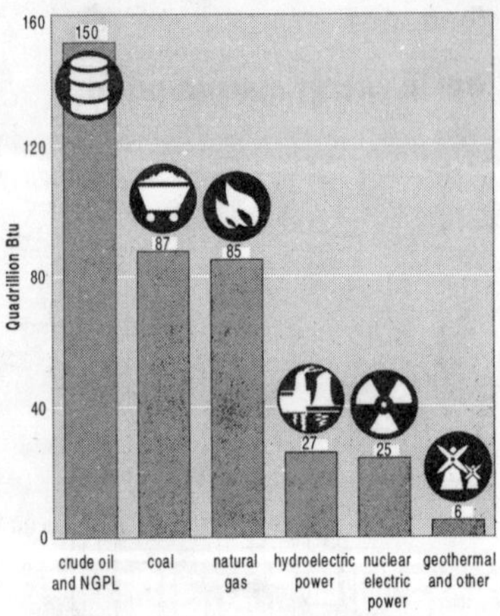

- **xylene (c6h4(ch3)2)**

Colorless liquid of the aromatic group of hydrocarbons made by the catalytic reforming of certain naphthenic petroleum fractions. Used as high-octane motor and aviation gasoline blending agents, solvents, chemical intermediates. Isomers are metaxylene, orthoxylene and paraxylene.

- **yellowcake**

A natural uranium concentrate that takes its name from its color and texture. Yellowcake typically contains 70 to 90 percent U_3O_8 (uranium oxide) by weight. It is used as feedstock for uranium fuel enrichment and fuel pellet fabrication.